The Value of Food Series

General Editors

Patty Fisher, B.Sc.(H & SS)
and

Arnold E. Bender, B.Sc., Ph.D., F.R.I.C., F.R.S.H., F.I.F.S.T.

P. Fisher and A. E. Bender: *The value of food* (1970)

The Value of Food Series

S. H. Cakebread: *Sugar and chocolate confectionery* (1975)
J. Scade: *Cereals* (1975)
J. W. G. Porter: *Milk and dairy foods* (1975)

Milk and dairy foods

J. W. G. PORTER
Head of Nutrition Department, National Institute for Research in Dairying

OXFORD UNIVERSITY PRESS
1975

Oxford University Press, Ely House, London W. 1
GLASGOW NEW YORK TORONTO MELBOURNE WELLINGTON
CAPE TOWN IBADAN NAIROBI DAR ES SALAAM LUSAKA ADDIS ABABA
DELHI BOMBAY CALCUTTA MADRAS KARACHI LAHORE DACCA
KUALA LUMPUR SINGAPORE HONG KONG TOKYO

ISBN 0 19 859431 3

© Oxford University Press 1975

All rights reserved. No part of this publication may be reproduced, stored in a retrieval system, or transmitted, in any form or by any means, electronic, mechanical, photocopying, recording or otherwise, without the prior permission of Oxford University Press

This book is sold subject to the condition that it shall not, by way of trade or otherwise, be lent, re-sold, hired out, or otherwise circulated without the publisher's prior consent in any form of binding or cover other than that in which it is published and without a similar condition including this condition being imposed on the subsequent purchaser.

Printed in Great Britain
by J. W. Arrowsmith Ltd., Bristol

Preface

Milk is of special significance in nutrition as the most nearly complete single food, but it is also important as the raw material from which a variety of nutritious products is made. These products may contain all or only some of the nutrients present in milk but each product can make a nutritionally significant contribution to our diets.

The purpose of this book is to provide information about the composition of milk and milk products, to relate composition and nutritive value, and to show how both are affected by the various manufacturing processes used in the dairy industry. I have tried to give sufficient information for teachers and students of home economics, household science, catering training, and allied subjects. I hope the book is also of interest to the general reader who would like to understand the scientific basis of practical nutrition.

J. W. G. P.

Contents

1.	The composition of milk	1
2.	The nutritive value of milk for man	7
3.	Milk in the diet	13
4.	Processed liquid milk	16
5.	Dried milk	25
6.	Infant foods	29
7.	Cheese	33
8.	Cream, butter, and margarine	42
9.	Ice cream	47
10.	Soured and fermented milks	51
11.	Legislation controlling milk and dairy foods	54
	Conclusion	57
	Questions	58
	Appendix	60
	Further reading	62
	Index	63

1 The composition of milk

Milk is the natural food of the new-born mammal, for whom it provides the sole source of nourishment during the period directly after birth. The fact that young animals grow and thrive when receiving a diet solely of milk might be taken to imply that their mother's milk is a complete food and that it supplies adequate amounts of all essential nutrients. This is not quite correct, since most milks contain insufficient iron and vitamin D for the needs of the young. One pint (0·57 l) of cow's milk supplies only one-thirtieth of a human infant's recommended intake of iron and less than one-hundredth of that for vitamin D. Nonetheless, milk is the most complete single food, and its value has been recognized since man's earliest days.

A number of species of large domesticated animals are used to provide milk for human consumption. In Britain, the cow is the most important dairy animal; in France and Italy, the sheep and the goat are also used; and in India and Egypt, the water-buffalo is the principal provider of milk. Animals of these species have been bred for centuries to provide quantities of milk far in excess of the amounts needed to nourish their young. A typical cow, for instance, will yield 850 gallons (3870 l) of milk during her period of lactation, which lasts for about 9 months, whereas her calf needs less than 50 gallons (230 l) to rear it to the stage when it can eat solid food and no longer requires milk.

The detailed composition of milk not only differs from one species of animal to another, but it varies quite widely within any one species and even between individuals of a breed or race of the same species.

THE CONSTITUENTS OF MILK

The milks of all animals consist largely of water (80–90 per cent) in which are dissolved or dispersed the proteins, the milk sugar (lactose), the minerals, and the water-soluble vitamins (Fig. 1.1). The milk fat is emulsified and is distributed throughout milk as minute droplets or globules, which may cluster together and rise to give a layer of cream when fresh milk is allowed to stand.

The characteristic 'milky' appearance of milk is due mainly to the dispersion of the milk proteins and calcium salts; the yellow colour of cream is due to the presence of carotene, an orange-yellow pigment that is converted to vitamin A (retinol) in the body.

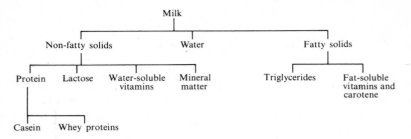

Fig. 1.1. The constituents of milk

The proteins of milk consist of casein and the whey proteins, mainly α-lactalbumin and β-lactoglobulin. Casein is a phosphorus-containing protein which is found only in milk and which forms the curd when milk is acidified or treated with rennet. The whey proteins remain dissolved in the liquid (whey) that drains from the curd.

THE COMPOSITION OF COW'S MILK

Most of the milk produced in Britain is taken to central dairies, where the contributions from individual farms are mixed together before the milk is processed. The composition of a typical sample of such bulk milk is shown in Tables 1.1 and 1.2. Differences in the composition of milk from cow to cow are to a large extent averaged out by bulking, but analyses of samples of bulk milk taken each month show that its composition varies in a regular manner throughout the year (Fig. 1.2). The reasons for this are discussed on p. 3.

TABLE 1.1
The major constituents of cow's milk (g per 100 g milk)

Water	87·6
Fat	3·8
Protein	3·3
casein	2·6
whey proteins	0·7
Lactose	4·7
Calcium	0·12
Non-fatty solids	8·7
Total solids	12·5

From the commercial standpoint the important features of milk composition are its content of fat, non-fatty solids (i.e. solids-not-fat, or SNF), and total solids, since the amounts of these constituents affect the yields of products such as butter, cheese, or concentrated milk that can

TABLE 1.2
Vitamins in cow's milk (per 100 g milk)

Retinol (vitamin A) (μg)	35	(summer)
	26	(winter)
Carotene (μg)	25	(summer)
	12	(winter)
Vitamin D (μg)	0·02	
Thiamin (μg)	45	
Riboflavin (μg)	180	
Nicotinic acid (μg)	80	
Pantothenic acid (μg)	320	
Vitamin B_6 (μg)	40	
Biotin (μg)	2·5	
Folic acid (μg)	6	
Vitamin B_{12} (μg)	0·35	
Vitamin C (mg)	2	

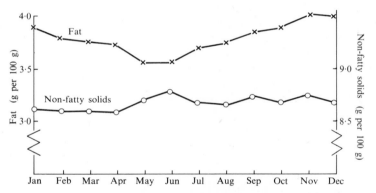

Fig. 1.2. Seasonal variation in the composition of milk

be obtained. The price the farmer receives for his milk depends on its content of SNF and total solids. Legal aspects of the compositional quality of milk are considered on p. 54.

Factors affecting the composition of cow's milk

1. *Feed.* Most of the constituents of milk are synthesized in the mammary gland from simpler substances derived from the food and passed to the gland in the blood-stream; e.g. lactose is synthesized from glucose, and milk proteins from amino acids. Milk fat originates partly from fat in the blood and partly from synthesis in the mammary gland (see p. 8). Other constituents, such as minerals and vitamins, are filtered out from the blood and passed into milk without being altered. To maintain milk secretion it is necessary

constantly to replace the supplies of raw material in the blood-stream by ensuring that the cow is properly fed. During the spring and summer cows live outdoors and eat grass, but during the winter they are normally kept indoors and given diets of hay or silage and concentrates (cereals and oil-seed meals).

Cows belong to the group of herbivorous animals called ruminants. Ruminants are characterized by their cloven hoofs and by their habit of chewing the cud. The group also includes sheep, goats, deer, and buffalo. Also, these animals have a digestive system that is specially adapted to deal with the bulky fibrous food that they eat. Ruminants have large stomachs, which are divided into four compartments. The food that they eat is stored in the first two of these compartments, the rumen and reticulum, which together form a large fermentation vat in which the food is digested by micro-organisms. Microbial digestion is helped by the animal chewing its cud, a process in which lumps of partially digested food are regurgitated from the rumen to the mouth to be moistened with saliva and broken up by chewing into smaller pieces, before being swallowed again. When the food has been digested by the micro-organisms it passes to the true stomach and then to the intestines, where the micro-organisms are themselves broken down by the animal's digestive enzymes and the nutrients are absorbed.

One of the principal functions of the rumen micro-organisms is to digest complex carbohydrates, such as cellulose, which cannot be digested by the digestive enzymes secreted by animals. Microbial digestion of cellulose and starch in the rumen results in the formation of large amounts of simple fatty acids (acetic, propionic, and butyric acids). Cellulose-digesting bacteria produce mainly acetic acid, whereas starch-digesting bacteria give rise predominantly to propionic acid. These fatty acids are absorbed directly from the rumen into the blood and are the important raw materials from which the main constituents of milk are synthesized. Acetic acid is used for the synthesis of milk fat, whereas propionic acid is the starting point for the synthesis of glucose (and hence lactose) and certain amino acids (and hence milk proteins).

To obtain maximum yields of milk containing a satisfactory content of fat and non-fatty solids, the cow's diet must provide both sufficient cellulose in the form of roughage and sufficient starchy foods to allow the rumen micro-organisms to synthesize adequate amounts of acetic and propionic acids. Reducing a cow's daily ration of hay from the normal amount of about 18 lb (8 kg) to 2 lb (1 kg) results in a markedly reduced synthesis of acetic acid and causes milk fat to fall from 3·6 per cent to below 2 per cent. A similar, though less extreme, effect often occurs in the spring when, as cows change from indoor feeding to grazing, the lower roughage content of their grass diet freqently brings about a depression in the fat content of milk (Fig. 1.2).

Ruminants are like other animals in that they require a dietary supply of vitamin A, which is provided by the carotene present in grass, silage, and hay. As grass contains much more carotene than hay, milk produced during the summer grazing season contains substantially more carotene and vitamin A than that produced during the winter (Table 1.2). The position is different with the vitamins of the B group, thiamin, riboflavin, nicotinic acid, pantothenic acid, vitamin B_6, biotin, folic acid, and vitamin B_{12}. All these vitamins are synthesised by the micro-organisms in the rumen, so that ruminants (unlike man, pigs, and poultry) do not need a supply of these vitamins in their food, and the concentrations of B vitamins in milk do not alter when the cow's diet is changed.

2. *Breed*. The values in Table 1.3 give some of the major constituents of representative samples of milk from two of the breeds of cow that produce much of the milk in the U.K. The compositions of the milk of other breeds, such as Ayrshire and Shorthorn, lie between those given for Friesian and Guernsey cows. There are about 3 million cows in the UK and about three-quarters of these are of the Friesian breed. These black-and-white cows give high yields of milk (up to 6 gallons (27 l) per cow per day), but their milk contains appreciably less fat and somewhat less non-fatty solids than the milk of other breeds. The Guernsey cow, like the Jersey and the South Devon, produces 'Channel Island' milk. Guernsey cows yield less milk than Friesians, but the milk has a higher content of fat and non-fatty solids.

TABLE 1.3
Constituents of the milk of Friesian and Guernsey cows (per 100 g milk)

	Friesian		*Guernsey*	
Water (g)	87·7		86·3	
Fat (g)	3·7		4·6	
Protein (g)				
casein	2·5		2·8	
whey proteins	0·7		0·8	
Lactose (g)	4·7		4·7	
Calcium (g)	0·12		0·13	
Retinol (μg)	37	(summer)	28	(summer)
	25	(winter)	23	(winter)
Carotene (μg)	24	(summer)	50	(summer)
	12	(winter)	24	(winter)
Non-fatty solids (g)	8·7		9·1	
Total solids (g)	12·4		13·7	
Energy content				
(kJ)	268		314	
(kcal)	64		75	

There are no appreciable differences due to breed in the levels of water-soluble vitamins in milk, but there is about twice as much carotene though slightly less retinol in the milk of Channel Island breeds than in others (see p. 11).

3. *Colostrum.* Colostrum, the secretion from the mammary gland for the first few days after parturition (the birth of young), differs markedly in composition from milk secreted thereafter. Cow's colostrum is sometimes known as beastings; it is characterized by a high content of whey proteins (up to 14 g per 100 g colostrum) and by a lower content of lactose; the fat content is similar to that of milk but the levels of fat-soluble vitamins are up to 5 times higher. The colostral whey proteins contain a large proportion of immune globulins, a special type of protein which provides the calf with antibodies that help to protect it against infections during the period of early life when it is unable to synthesize its own.

THE COMPOSITION OF THE MILK OF OTHER MILCH ANIMALS

The values in Table 1.4 show the composition of buffalo's, goat's, and ewe's milk. Goat's milk closely resembles cow's milk, but ewe's milk and buffalo's milk are much richer in fat and slightly richer in protein than cow's milk. These milks are always pure white in colour, because none of them contains carotene.

TABLE 1.4
Some constituents of milk of different species (per 100 g)

	Buffalo	*Goat*	*Ewe*
Protein (g)	3·8	3·3	5·6
Fat (g)	7·5	4·5	7·5
Lactose (g)	4·9	4·4	4·4
Calcium (g)	0·19	0·13	0·20
Retinol (μg)	65	40	65
Carotene (μg)	0	0	0
Thiamin (μg)	50	50	70
Riboflavin (μg)	100	120	500
Folic acid (μg)	—	0·2	0·2
Vitamin B_{12} (μg)	0·3	0·1	0·3
Energy content			
(kJ)	430	300	450
(kcal)	102	71	108

2 The nutritive value of milk for man

Foods are needed for the production of heat, for muscular work, and for body construction and repair. A complete diet supplies all the nutrients needed for these purposes. We have seen that milk is almost a complete food in itself since it contains both energy-supplying nutrients (fat and carbohydrate) and body-building nutrients (proteins and minerals) and also adequate amounts of almost all the vitamins necessary for the proper functioning of the biochemical processes which are carried out in our bodies and which are essential for life.

ENERGY-SUPPLYING NUTRIENTS

Carbohydrates
These substances are composed chemically of carbon, hydrogen, and oxygen; those occurring in food include the sugars (mono-, di-, and trisaccharides) and starch and cellulose (polysaccharides). Only monosaccharides can be absorbed from the intestine into the blood, so that other carbohydrates have to be broken down to their constituent monosaccharides by digestive enzymes secreted by the pancreas and small intestine (cellulose can only be broken down by micro-organisms, see p. 4). Each gram of absorbed sugar provides the body with 4 kcal (16·7 kJ) of energy.

Lactose, the milk sugar, is a disaccharide and is broken down during digestion by the enzyme lactase to the monosaccharides, glucose and galactose. Human infants, like all young mammals, secrete the high levels of lactase necessary to enable them to digest lactose in their milk, but after weaning the secretion of lactase falls. In individuals of some human races, particularly among coloured peoples in parts of Africa and south-east Asia, lactase secretion stops altogether after weaning. The resulting lactase deficiency prevents such individuals from digesting lactose and may cause them to suffer from diarrhoea if they drink a large quantity of milk.

Fats
Fats are a more concentrated source of energy than carbohydrates and each gram provides 9 kcal (37·7 kJ). They consist of triglycerides in which glycerol is combined with 3 fatty acids which may be the same or may be any of a dozen or more different ones.

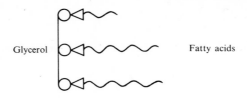

Diagram of a triglyceride

Fats consist mainly of mixtures of triglycerides, and they occur widely in foods of both animal and vegetable origin. Each naturally occurring fat contains its own particular pattern of triglycerides with their characteristic component fatty acids. These fatty acids have a chain of 2–22 carbon atoms (shown by the wavy lines in the diagram). Most fatty acids belong either to the series of saturated fatty acids, in which each carbon atom carries its full quota of hydrogen atoms, or to the group of unsaturated fatty acids, in which one or more pairs of hydrogen atoms are missing. Butyric acid with 13 carbon atoms, palmitic acid with 16, and stearic acid with 18 are typical saturated fatty acids; whereas oleic acid with 18 carbon atoms, and the two fatty acids that are essential in our diet—namely, linoleic acid with 18 carbon atoms and arachidonic acid with 20—are examples of unsaturated fatty acids.

Milk-fat triglycerides contain a large number (over 100) of different fatty acids, each of which belongs to one of three groups: short-chain with 2–8 carbon atoms per molecule, medium-chain with 10–14 carbon atoms, and long-chain fatty acids with 16 or more carbon atoms.

The triglycerides of milk fat are built up from fatty acids containing 16 and 18 carbon atoms derived from the triglycerides of blood plasma and from fatty acids with shorter carbon chains synthesized in the mammary gland.

Milk fats from ruminant animals—cows, sheep, goats, and buffaloes—have a high content of short-chain fatty acids with 4–8 carbon atoms. This is because in these animals the shorter-chain fatty acids are built up in the mammary gland from acetic acid (which has 2 carbon atoms). Human milk fat and the milk fat of other non-ruminant animals such as pigs and dogs do not include any fatty acids with less than 10 carbon atoms; in these species glucose (with 6 carbon atoms) is the starting point for the synthesis of the shorter-chain fatty acids.

The fatty acids in the milk-fat triglycerides are not haphazardly distributed but they are arranged so that most of the triglycerides contain 1 fatty acid with less than 10 carbon atoms, and 2 fatty acids with more. This arrangement is believed to contribute to the ease with which milk fat is digested as compared with some other fats, particularly those with a large proportion of triglycerides each containing 3 fatty acids with 18 carbon atoms.

BODY-BUILDING NUTRIENTS

Proteins

Proteins differ from fats and carbohydrates in that they contain about 16 per cent of nitrogen. The protein content of a food is usually determined by measuring its nitrogen content and multiplying the result by 100/16 (6·25) to give an estimate of the amount of 'crude protein' present. Milk proteins contain slightly less than 16 per cent nitrogen so that in their determination the nitrogen content is often multiplied by 6·38.

All proteins are built up from 20 different amino acids. During digestion, food proteins are broken down to their constituent amino acids by the enzymes of the digestive juices. The amino acids are then absorbed into the blood-stream and carried to the tissues, where they are rebuilt into the particular pattern required in the proteins of muscle, skin, or body organ. The body can make 12 of the amino acids it needs, but the other 8 must be provided ready-made in the diet. These are called 'essential amino acids'. The increased requirements for growth make 2 other amino acids—arginine and histidine—essential for children (see Table 2.1).

TABLE 2.1
Amino acids

Essential	*Non-essential*
Leucine	Alanine
Isoleucine	Glycine
Lysine	Serine
Methionine	Proline
Valine	Hydroxyproline
Threonine	Cystine
Tryptophan	Cysteine
Phenylalanine	Tyrosine
	Glutamic acid
Arginine for children	Aspartic acid
Histidine	

Food proteins differ in quality, i.e. in their ability to supply the correct amounts of essential amino acids required by the body. The proportion of a dietary protein that is retained in the body for the synthesis of new tissue is a measure of its usefulness and is termed its 'biological value'.

The proteins of egg are well endowed with all the essential amino acids and have the highest possible biological value (BV) of 1·0. Milk proteins are also of very good quality. The whey proteins (BV 1·0) are slightly superior to casein (BV 0·80). Casein has a lower BV than the whey proteins because it has a slight deficiency of methionine plus cystine. However, the whey proteins have a slight surplus of these amino acids, so that when casein and the whey proteins are combined in milk they complement one another, and the BV of milk protein depends on the relative amounts of the proteins that it contains. Cow's milk protein contains 4 parts casein to 1 part whey proteins and has a BV of 0·88, whereas human milk protein contains equal amounts of casein and whey proteins, and the latter can provide enough cystine to raise the BV to 1·0.

Both casein and the whey proteins have a surplus of the essential amino acid, lysine. This means that in a mixed diet milk proteins can be of particular value in enhancing the value of other proteins, such as those of cereals, which have a low content of lysine. Everyday examples of complementation of this kind include eating bread with cheese (see p. 40), drinking milk with biscuits, and adding milk to breakfast cereals. The proteins of breakfast cereals have a BV of about 0·50, but the addition of ¼ pint (140 ml) of milk to a 1¾ oz (50 g) helping of wheat- or cornflakes raises the BV to 0·85. The total protein content of such a helping is the same as that supplied by an egg and 1¾ oz (50 g) of bacon, the proteins of which also have a BV of about 0·85.

In a similar way the quality of the proteins of batters and doughs can be improved by incorporating dried whole or skimmed milk. Optimal complementation would require the addition of about 1 part of dried milk to 2 parts of flour, but in practice not more than 1 part to 8 parts can be used without affecting the texture. The British 'milk loaf' contains 6 per cent added milk solids.

Minerals

The body requires a number of mineral elements, all of which must be obtained from food. Minerals are necessary constituents of body cells and fluids and two elements, calcium and phosphorus, are also needed in relatively large amounts for the formation of bones and teeth. Some elements are needed for special purposes, e.g. iron is a constituent of haemoglobin, the red pigment which gives blood its colour and which is necessary for the transport of oxygen from the lungs to the tissues. Phosphorus (as phosphate) and most of the minerals that are needed in only small amounts are widely distributed in foods, so that intake from normal diets are sufficient for the body's requirements. There are fewer good sources of calcium and iron. For this reason in the UK, and in many other countries, white flour is fortified with calcium and iron to help to ensure that the diet supplies adequate quantities of these elements.

Milk is an exceptionally good source of calcium: half a pint (0·28 l) of milk provides a child with half and an adult with two-thirds of his daily need. Cheese is also an excellent source, since nearly all the calcium in milk is retained in the curd during the making of all but cottage cheese.

Milk and cheese are poor sources of iron.

VITAMINS

Vitamins are organic compounds that are required in minute amounts for life and health. Four of the vitamins—namely, A, D, E, and K—are found mainly in fatty foods and are called fat-soluble vitamins. The other vitamins are soluble in water and include ascorbic acid (vitamin C) and the vitamins of the B group. The B vitamins likely to be in short supply in our diet and therefore of most practical interest are thiamin (B_1), riboflavin (B_2), nicotinic acid, folic acid, and vitamin B_{12}.

Fat-soluble vitamins

Vitamin A. In the body vitamin A is necessary for growth, for normal vision, and for maintaining in a healthy condition the moist surface tissues such as the lining of the throat and respiratory tract.

Vitamin A is present in foods in two different forms. The vitamin itself, retinol, occurs only in certain animal fats such as those of milk and eggs. On the other hand, plants and foods of vegetable origin contain an orange substance, carotene, which is converted to retinol in the walls of the intestine of animals. Some animals, including man and cattle, can also absorb some unchanged carotene, and their blood and milk contain both carotene and retinol. In most animals 1 mg of carotene yields about 0·16 mg of retinol, but in cows and other ruminant animals the conversion is often much less efficient. For instance, cows grazing pasture, which is a rich source of carotene, may ingest 4 g of carotene per day but only secrete into milk a few milligrams of retinol. Different breeds of cattle show marked differences in their utilization of carotene. The Channel Island breeds take up more carotene from their food and produce yellower milk fat containing much more carotene though a little less retinol than do other breeds (see Table 1.3).

Vitamin D. Vitamin D is necessary for the absorption of calcium. The natural form, vitamin D_3 or cholecalciferol, occurs in foods and can be made in the skin by the action of ultraviolet light from sunshine. The synthetic form, vitamin D_2 or ergocalciferol, is prepared by the action of ultraviolet light on ergosterol, a substance that occurs in plants. Milk contains both vitamins D_2 and D_3; vitamin D_2 derives from hay, in which it is formed whilst the grass is drying in the sunshine during haymaking, and vitamin D_3 from the action of sunshine on the cow. Milk contributes only small quantities of vitamin D to the human diet.

Babies fed on cow's milk need a supplement of vitamin D, which can be provided by cod-liver oil or one of the proprietary preparations.

The vitamin D content of milk can be increased by irradiating it with ultraviolet light or, more simply, by adding to it an emulsified preparation of the vitamin. Vitamin D is added to evaporated milk and to other milk products that may be used for feeding babies.

Water-soluble vitamins

B vitamins. These vitamins form part of the tissue-cell enzymes that act as catalysts for the many chemical reactions carried out in the body. Three of them, thiamin, riboflavin, and nicotinic acid, are primarily concerned with the oxidation of energy-supplying nutrients in body cells. Two others, folic acid and vitamin B_{12}, play an important part in the formation and development of red blood cells.

In the cow and other ruminant animals all these vitamins are synthesized by micro-organisms in the rumen, and ruminants do not need a supply of B vitamins in their diet. For this reason the level of these vitamins in their milk varies hardly at all, whereas in the milk of simple-stomached animals (e.g. in human milk) the levels of B vitamins depend fairly directly on the vitamin intake from food. Typical values for the B vitamin content of milk were given in Table 1.2.

Milk is particularly important as a source of riboflavin in the diet: half a pint (0·28 l) of milk provides a child with half and an adult with two-thirds of his daily need. The riboflavin of milk is slowly destroyed by light, and much of the vitamin can be lost if milk in a clear glass bottle is left in bright sunshine for several hours.

Ascorbic acid (vitamin C). The main function in the body of vitamin C, the anti-scurvy vitamin, is to assist in the formation of connective tissue. Freshly drawn cow's milk contains about 2 mg of ascorbic acid per 100 g of milk, but the vitamin is readily destroyed by the effects of light and oxygen so that the vitamin C content of milk as consumed is usually very much lower and probably not more than 1 mg per 100 g. As with the B vitamins, the vitamin C content of human milk varies, depending on the mother's intake of the vitamin. Cows and most other animals do not need a dietary supply of vitamin C as they synthesize it in their body tissues and the level in their milks does not vary.

3 Milk in the diet

Apart from its essential value as a sole food in the feeding of infants (Chapter 5), milk and dairy products continue to make an increasing contribution to the typical mixed diet eaten in the UK. The major part of this contribution is from liquid milk.

Milk production and consumption have increased steadily during the last 100 years, and our average annual consumption per head of about 255 pints (145 l) is nearly 50 per cent greater than before the 1939–45 war and about 4 times as much as it was a century ago. Improvements in methods of livestock farming, in transport facilities for distributing milk, and in dairy practice have all helped to make good-quality milk widely available. At the same time there has been increased awareness of the importance of diet and nutrition to the health of growing children and adults.

The particular value of milk in improving children's diets was clearly established by tests carried out in the 1920s by Dr. Corry Mann and by Sir John Boyd Orr (later Lord Boyd Orr). This work was followed by the introduction of the first School Milk Scheme in 1927; under this scheme children were provided with $\frac{1}{3}$ pint (0·19 l) at 1 (old) penny per bottle. This scheme grew voluntarily, with the Government subsidizing the cost for elementary school children. The 1944 Education Act provided milk free of charge for children attending school. This scheme remained essentially unaltered until 1968, when free milk was withdrawn from secondary schools. In 1971, the scheme was again modified, and at present children in primary schools receive free milk until the end of the summer term after they reach the age of seven.

In adult diets milk is not necessarily taken as a drink, but may also be taken by adding it to breakfast cereals, tea, or coffee and by using it in cooked dishes such as soups, sauces, and milk puddings. Our average daily consumption per head is about $\frac{2}{3}$ pint (0·38 l), though, of course, more is drunk by children and less by adults. A child drinking 1 pint (0·57 l) of milk a day will be provided with all its daily needs for riboflavin, vitamin B_{12}, and calcium, and with a large proportion of the protein, thiamin, folic acid, and vitamin A (see Table 3.1).

When the contributions from all dairy products are added together (Table 3.2) it is evident that they supply about a fifth of the energy, a quarter of the protein, a third of the riboflavin and vitamin A, and nearly two-thirds of the calcium consumed in the UK.

TABLE 3.1
Percentage contribution of 1 pint (0·57 l) of Friesian milk to recommended daily nutrient intakes of a 5–7-year-old child

	Recommended intake	Percentage from milk
Protein (g)	45	40
Thiamin (mg)	0·7	35
Riboflavin (mg)	0·9	100
Nicotinic acid (mg)	10	5
Folic acid (μg)	100	35
Vitamin B_{12} (μg)	1·5	100
Ascorbic acid (mg)	20	30
Vitamin A (retinol equivalents, μg)	300	60
Vitamin D (μg)	10	1
Calcium (mg)	500	100
Iron (mg)	8	0
Energy content		
(kJ)	7500	20
(kcal)	1800	20

TABLE 3.2
Percentage contribution of dairy products to total amount of certain nutrients consumed in UK

	Energy	Protein	Calcium	Riboflavin	Vitamin A (retinol equivalents)
Liquid milk	11·0	18·1	46·5	38	12·5
Dried milk	0·2	0·3	0·8	6	0·6
Other milk and cream	1·1	1·3	3·1	2	1·4
Cheese	2·2	5·3	11·4	–	4·8
Butter	6·6	0·1	0·3	–	16·8
Total dairy products	21·1	25·1	60·6	46	36·2

Based in part on *United Kingdom dairy facts and figures* (1974). United Kingdom Federation of Milk Marketing Boards.

Although dairy products are of value as a source of energy-supplying nutrients, their particular importance as food is measured much more adequately by this contribution of high-quality protein, calcium, and certain vitamins.

As already mentioned, milk proteins are not only valuable in themselves but also through the way in which they can enhance the value of other proteins in the diet. The proteins of cereals and of most foods of vegetable origin are of markedly lower biological value than milk proteins, mainly because they have a low content of the essential amino acid, lysine. Milk proteins are thus able to complement other proteins and to improve their usefulness. This has been clearly shown in experiments both with rats and human subjects, in which bread and cheese were given first together, then on different days (see p. 40 and Table 7.3). The same effect has been found with milk and potatoes, and milk and cereals.

4 Processed liquid milk

THE NEED FOR HEAT-TREATMENT

The composition of milk makes it not only an excellent food for man but also an ideal medium for the growth of bacteria and other micro-organisms. Milk can be drawn from the udder of a healthy cow by special methods that ensure that it is sterile and contains no micro-organisms. However, normal milking procedures usually yield milk that is infected with a small number of micro-organisms, some of which may be harmful (pathogenic) to man. Micro-organisms grow and multiply rapidly in milk. As they grow they produce acid, particularly lactic acid, which causes the milk to develop a characteristic sour taste and eventually, when enough acid has been produced, to clot.

The organisms causing tuberculosis and brucellosis (undulant fever) may be present in the milk of cows suffering from these diseases. Other organisms, such as *streptococci* and *staphylococci,* may pass into milk if the cow has an infected udder. All cattle in the UK are now tested regularly to ensure that animals infected with tuberculosis are weeded out and in a large part of the country similar tests are being made to detect and slaughter animals infected with brucellosis. After it leaves the udder, milk may also be infected from utensils or people, and outbreaks of disease or food-poisoning have been caused in this way.

To ensure that milk sold for human consumption is a safe product of good keeping quality it is necessary to arrange a system of handling and processing that destroys all the harmful micro-organisms and reduces the population of other micro-organisms. Since bacteria multiply more rapidly in warm than in cold milk, the first step in minimizing the rate of spoilage after it leaves the cow's udder at body temperature is to cool it at the farm to around 45 °F (7 °C). The resulting cold milk is then transported to central dairies (often called creameries) for further processing. The most effective means of destroying bacteria in milk is to heat it for a sufficient time at a temperature just high enough to kill all pathogenic organisms. This is the principle of pasteurization.

The conditions chosen for pasteurizing are not severe enough to kill all micro-organisms, but in practice it is found that besides the pathogens most of the other bacteria responsible for causing milk to sour are also killed. The mild heating conditions used in pasteurizing cause little loss of nutrients, but it is also important that pasteurizing does not affect the flavour of milk and does not cause any reduction in the cream layer which forms when the milk is allowed to stand. Overheating during pasteurization prevents the globules of fat forming the clusters which rise to the surface to form the cream layer or cream line; it does not, of course, reduce the amount of fat present in the milk.

Pasteurized milk is immediately filled into a clean container that can be sealed against the risk of recontamination by micro-organisms. Most milk is still sold in the familiar pint glass bottles, but increasing use is being made of cartons and plastic bags or bottles.

DISTRIBUTION AND STORAGE

Milk should be protected from light to prevent losses of ascorbic acid and riboflavin. When distributed in glass bottles they should be delivered in covered vans, brought into the house as soon after delivery as possible, and placed in a refrigerator or other cool, dark place. A pint bottle of milk left to sit on the doorstep in bright sunshine for an hour will have lost most of its ascorbic acid and up to half its riboflavin (see p. 12).

One of the advantages of brown-glass bottles, cartons, and plastic containers is that they restrict the action of light, but there is still the same need to keep the milk cool.

PASTEURIZED MILK

Processing

Pasteurization consists of heating milk below the boiling point but at a temperature sufficiently high to kill pathogenic organisms and reduce the numbers of the others so as to enable the product to be safely transported, distributed, and consumed as liquid milk. Pasteurization destroys all the pathogenic organisms and about 99 per cent of the other bacteria in milk. Most of the milk produced in the UK is pasteurized by the high-temperature short-time (HTST) process, though a few smaller dairies still use the older 'holder' process.

HTST pasteurization

A typical modern HTST plant will process 10 000 litres of milk per hour, i.e. enough to fill 16 000 pint bottles.

Cold, raw (i.e. unpasteurized) milk from a storage tank is pumped through a heat exchanger, where it is initially heated by hot milk that is being cooled down. The milk is then filtered through a cloth filter and further heated by hot water to a final temperature of 161 °F (71·7 °C). The milk then passes through a holding tube which is so designed that, in flowing through it, all the milk is held at this final temperature for at least 15 seconds. A flow-diversion valve at the outlet of this tube is controlled by the temperature of the milk flowing through it. If the temperature falls below 161 °F (71·7 °C) the valve operates and the milk is returned to the storage tank for reprocessing. Milk that has been properly heated is allowed to flow back through the heat exchangers, and in doing so it is itself cooled whilst warming up the incoming milk. It then passes through a final cooling section in which it is cooled by chilled water to a temperature of not more than 50 °F (10 °C).

The heat exchangers, like all the other metal parts of the HTST plant coming into contact with milk, are made of stainless steel. They consist of a number of plates spaced apart by rubber seals to form a narrow chamber between each pair of plates. The liquid being heated and the liquid being cooled flow through alternate chambers.

Holder pasteurization
In the holder or batch method, milk is pasteurized in individual batches of 50–300 gallons (200–1500 l). The pasteurizer consists of a large stainless-steel tank in which the milk is stirred during heating, holding at 145–150 °F (62·8–65·6 °C) for 30 minutes and then cooling. This tank is surrounded by an outer casing and the space between the casing and the tank forms a jacket through which hot water or steam is passed to heat the milk, and then cold water to cool it. The milk is then passed through a cooler and filled into bottles.

Effect of pasteurization on nutritive value
Pasteurization of milk either by the HTST or holder process gives a product unchanged in flavour that will keep for several days if kept cool.

The effects on nutritive value are small. The major nutrients—proteins, fat, carbohydrate, minerals, and most of the vitamins—are unaffected. The only appreciable loss is of about half the vitamin C and of about 10 per cent of the thiamin and vitamin B_{12} originally present. By itself vitamin B_{12} is stable to heat and should therefore withstand the pasteurizing process without loss. The small loss that occurs is due to interaction of vitamin B_{12} with the products formed during the destruction of vitamin C.

BOILED MILK

Boiling kills most of the micro-organisms in milk but it is a more drastic heat-treatment than pasteurization. It alters the flavour and it causes greater losses of vitamin C, thiamin, and vitamin B_{12}, and also loss of folic acid. The main losses, however, are in the skin which tends to form on the top of the milk and in the deposit of protein and calcium which adheres to the bottom of the pan. These losses may amount to one-sixth of the protein and calcium and one-fifth of the fat. Losses are minimized if milk is stirred whilst being rapidly heated to boiling and then quickly cooled.

Milk that has started to sour cannot be preserved by boiling since heating such milk causes the protein to coagulate.

HOMOGENIZED MILK

The process of homogenization consists in forcing milk heated to about 139 °F (60 °C) through a very small tube at high pressure; this breaks up the fat globules into very small droplets that remain suspended in the milk and do not float to the top; little or no cream layer is formed when the milk is allowed to stand.

Raw milk contains lipase, an enzyme that attacks milk fat, releasing free fatty acids which give the milk a bitter or rancid taste. So long as the fat is contained within the globules it cannot be attacked by lipase. Since homogenizing breaks up the globules and allows lipase to act, it is necessary to destroy the enzyme, either immediately before or after homogenizing. This is done by heat-treating the milk, either by pasteurizing or sterilizing.

Homogenization does not affect the keeping quality or nutritive value of milk.

STERILIZED MILK

Sterilization is a process that ensures the destruction of *all* micro-organisms. To sterilize milk it must be heated to a temperature substantially higher than 212 °F (100 °C). This is done either by ultra-high-temperature (UHT) sterilization, a process in which a continuous flow of milk is rapidly heated to 304 °F (150 °C), held at this temperature for 1 second, and then quickly cooled, or by in-bottle sterilization, a process in which milk is filled into bottles which are then heated to around 249 °F (120 °C) for ¼–1 hour.

Sterilized milk, unlike pasteurized milk, can be kept at room temperature for several months in an unopened container, without souring or spoiling due to the growth of bacteria. Once the container has been opened the milk can be re-contaminated by micro-organisms and its keeping properties become the same as those of pasteurized milk.

UHT sterilization

Milk can be sterilized by continuous processes by either direct or indirect ultra-high-temperature (UHT) procedures in which it is heated to 276–304 °F (135–150 °C) for at least a second. Milk is always homogenized before UHT sterilization. The UHT procedures are based on the discovery that higher processing temperatures with shorter holding times gave a product in which all bacteria had been killed but in which there was much less change in colour and flavour than was possible with the in-bottle sterilizing process.

Various designs of UHT plant are used. In the direct heating process the required temperature 278 °F (138 °C) is achieved by passing milk through a heat exchanger, using steam at a pressure of about 2·25 atm (225 kPa) for the final stage of heating. In the direct process the milk is heated with steam, either by injecting steam directly into milk or by blowing the milk through a fine nozzle into a tank filled with steam. The steam that condenses heats the milk almost instantaneously to about 304 °F (150 °C). The milk is then cooled in a chamber in which the pressure is maintained below that of the atmosphere by means of a vacuum pump. This causes the surplus water in the milk from the condensed steam immediately to evaporate.

UHT sterilized milk is filled into special carton containers since there is as yet no satisfactory way of filling it into glass bottles without risk of contamination with micro-organisms. The commonly used Tetra-Pak cartons are plastic-coated and lined with aluminium foil, which ensures that the milk is kept in the dark.

The UHT process is also used as a preliminary stage in the preparation of in-bottle sterilized milk.

Effect of UHT sterilization and storage on nutritive value. Sterilization of milk by UHT processes has the same negligible effects on the nutritive value of milk as those caused by pasteurization. However, unless all the oxygen dissolved in the milk is removed during UHT treatment, further and more serious losses of certain vitamins may occur during storage after the milk has been filled into cartons. These losses of vitamin C, folic acid, and vitamin B_{12} do not occur in milk that has been treated by the direct heating process because all the oxygen is removed from the milk during cooling. To prevent losses from milk treated by the indirect process it is necessary to introduce a simple modification to the procedure to ensure that oxygen is removed.

TABLE 4.1
Losses of vitamins from UHT milk on keeping

	Oxygen (present +; absent −)	Days kept	Vitamin remaining (percentage of originally present)
Vitamin C	+	7	0
	−	60	80
Folic acid	+	14	0
	−	60	80
Vitamin B_{12}	+	60	40
	−	60	75

In-bottle sterilization
To prepare milk for the traditional in-bottle sterilization procedure it is first warmed to about 108 °F (43 °C) and filtered. It is then homogenized

and passed to a bottle-filling machine and filled into clear glass bottles with a narrow neck (1 in, 25 mm). Immediately after filling the bottles are capped with seals similar to those on beer bottles. These seals are able to retain the pressure developed during the subsequent heat-treatment, and the vacuum formed on cooling.

The sealed bottles of milk are heated to sterilizing temperatures by steam under pressure, either in a batch or a continuous process. In the batch process several hundred bottles at a time are loaded into a large tank (rather like a giant pressure cooker) which can be sealed before steam is introduced to raise the temperature to around 249 °F (120 °C). Heating is continued for 30 minutes to ensure that milk inside each bottle reaches a temperature of at least 240 °F (115 °C) for 5 minutes. In the continuous process the bottles of milk are loaded mechanically into a conveyor which carries them through a water seal into a sterilizer, where they are heated as in the batch process, and out again through another water seal.

The relatively drastic heat-treatment used in these methods imparts to milk a marked 'cooked' flavour and a richer colour and texture than does pasteurization. Perhaps for these reasons it is preferred by some consumers.

During the last few years some dairies have modified their procedure for preparing in-bottle sterilized milk. In the modern procedure milk is homogenized, sterilized by the UHT process, and then filled into bottles. Since the bottles are not sterile and cannot be filled without risk of bacterial contamination a further sterilization is necessary, but this does not require such severe heating as is used in the older process. The usual procedure is to sterilize the bottles at 230–234 °F (110–112 °C) for 15–20 minutes. Milk sterilized in this way has been less severely heated than by the older procedure and has a less marked cooked flavour.

Effect of in-bottle sterilization on nutritive value. In-bottle sterilization involves a relatively drastic heat-treatment which causes a slight overall reduction in the nutritive value of milk. The biological value of the proteins is slightly reduced and about one-third of the thiamin and half of the vitamin C, folic acid, and vitamin B_{12} are destroyed.

Since the milk may be kept in clear glass bottles for periods of several weeks, it is important to keep the bottles in the dark to prevent light causing loss of riboflavin and, after continued exposure, of retinol.

DIFFERENT TYPES OF MILK

Standardized milk

Standardized milk is milk in which, without altering any of the other constituents, the fat content is adjusted to a pre-determined value. The fat content can be raised by adding cream or lowered either by partial skimming or by adding skim milk.

At present, milk in the UK is not standardized but under EEC regulations to come into force in 1975 there will be three main categories: whole milk, containing not less than 3·5 per cent fat; half-skimmed milk, containing 1·5–1·8 per cent fat; and skim milk, containing less than 0·3 per cent fat.

Skim milk

Skim milk, or separated milk, is milk from which almost all of the fat has been removed. (The amount of fat left is usually about 0·1 per cent.) Skim milk contains no vitamin A and is not suitable for babies unless they are given supplements of vitamins A and D, which can be provided by cod liver oil or one of the proprietary preparations. For all other purposes it is an excellent food. It is claimed to be of value in slimming, since it has an energy content only about half that of the same volume of whole milk.

Filled milk and toned milk

These are two liquid milk products that use dried skim milk as an ingredient.

Filled milk is the term used to describe 'whole milk' made from reconstituted skim milk into which fat or oil other than milk fat has been incorporated.

Toned milk is the term used, particularly in Asian countries, to describe milk to which reconstituted skim milk has been added to reduce the fat content and at the same time to increase the content of non-fatty solids. In India, toning is a valuable means of augmenting locally produced supplies of buffalo milk, which has a high fat content of about 7·5 per cent. Three parts of reconstituted skim milk can be added to each part of buffalo milk to give a toned milk containing nearly 2 per cent fat.

Concentrated milks

These products are prepared by concentrating whole milk or skim milk to about one-third volume by the removal of water. The growth of microorganisms is prevented either by heat-treatment or by adding sugar. These concentrated milks are usually filled into cans. They keep well and have proved popular products with a wide range of uses.

The term 'evaporated milk' refers to products which have been heat-treated to prevent bacterial spoilage. Most of the evaporated milk manufactured in the UK is prepared from whole milk.

The terms 'condensed' and 'sweetened' condensed milk refer to products to which sufficient sugar (sucrose) has been added to preserve them. Most of the condensed milk manufactured in the UK is prepared from skim milk.

Evaporated milk. To ensure the manufacture of a uniform product the chemical composition of the raw milk is adjusted so that the ratio of fat to total solids is exactly 1:2·44. This value is close to that normally found in bulk milk and if adjustment is necessary it is made by appropriate addition of cream or skim milk to the original milk.

Milk entering the evaporating plant is first heated to around 193 °F (95 °C) and held at this temperature for 10 minutes to kill most bacteria. It is then run into the evaporator, which consists of a tall tube-shaped tank heated by steam and connected through a condenser to a vacuum pump. The action of the pump reduces the pressure, causing the milk to boil vigorously. The rates of heating and pumping are adjusted so that water is rapidly distilled from the milk at a temperature of about 120 °F (50 °C). After concentration the milk is homogenized and sealed into cans which are then sterilized in a steam autoclave at 240 °F (115 °C) for 15 minutes.

Condensed milk. The procedure is similar in principle to that used for the preparation of evaporated milk. Skim milk is used for the preparation of condensed skim milk and milk with a fat to total solids ratio of 1:2·44 for the preparation of condensed whole milk. The milk is pre-heated at 176 °F (80 °C) for 15 minutes, and introduced into the evaporator, where it is mixed with the required quantity of sugar and the mixture is boiled under reduced pressure to remove water. The amount of sugar added, 15–17 per cent by weight, is such that its concentration in the water of the condensed milk is about 62 per cent.

After concentration the milk is cooled to bring about crystallization of lactose in small crystals that do not cause the product to have a gritty texture. It is then filled into cans or tubes which are sealed.

Nutrient content of concentrated milks. Table 4.2 shows that the contents of total milk solids in evaporated and condensed whole milk are about 2·5 times those in the original milk. Thus to re-make milk 1 part of evaporated milk should be added to 1·5 parts of water.

Heat-treatment during the preparation of evaporated milk causes similar losses of nutrients to those caused during the older procedure for in-bottle sterilization; i.e. losses of about half the contents of thiamin, vitamin C, and folic acid, nearly complete loss of vitamin B_{12}, and a slight reduction in the biological value of the milk proteins. Most manufacturers add vitamin D to evaporated milk, particularly when it is likely to be used for feeding to babies.

Heat-treatment during the preparation of condensed milk is much less severe and causes only the same negligible loss of nutrients as does pasteurization.

Condensed milks contain a large amount of sugar which increases the energy content very considerably, but it significantly reduces the proportion of protein in the total solids. For this reason, and also because it contains no fat and hence no vitamins A and D, condensed skim milk is not a suitable food for babies.

Keeping quality of concentrated milks. Evaporated and condensed milks tend to thicken during prolonged storage but they will remain palatable for a year or more, particularly if kept in a cool place, i.e. at temperatures below 68 °F (15 °C). Such storage causes little change in the biological value of the proteins, though further progressive loss occurs of water-soluble vitamins. Storage at higher temperatures causes more rapid loss of vitamins, increases the rate at which the milk proteins tend to solidify and leads to impairment of their biological value due to loss of lysine.

TABLE 4.2
Composition and energy contents of evaporated and condensed milks (per 100 g)

	Evaporated		Condensed	
	Whole milk	Skim milk	Whole milk	Skim milk
Water (g)	68·5	80·0	25·0	29·0
Fat (g)	9·2	0·2	9·2	0·3
Protein (g)	8·4	7·4	8·4	9·6
Carbohydrate (g)	12·0	10·7	55·4	58·8
Calcium (g)	0·3	0·3	0·3	0·3
Vitamin A (retinol equivalents, μg)	80	2	80	3
Vitamin D (μg)	0·1†	0	0·1	0
Thiamin (μg)	65	65	100	110
Riboflavin (μg)	450	400	450	500
Folic acid (μg)	8	8	15	16
Vitamin B_{12} (μg)	0	0	0·8	0·8
Vitamin C (mg)	2	2	4	4
Total milk solids (g)	31	20	31	26
Total solids (g)	31	20	74	71
Energy content				
(kJ)	705	310	1410	1150
(kcal)	168	74	336	275

† Or 3μg when a supplement is added.

5 Dried milk

Drying has long been one of the simplest and most effective ways of preserving food, as micro-organisms cannot multiply in the absence of water. The preparation of dried milk provides a means of conserving milk nutrients in a form which is more cheaply and easily transported and stored than is concentrated liquid milk. A good-quality dried milk powder can be reconstituted to give a liquid that is virtually indistinguishable from milk.

Dried products are made from whole milk, milk from which part of the fat has been removed, and from skim milk, and also from by-products such as buttermilk and whey. The usual method of removing water is by heating and two processes are commonly used: roller-drying and spray-drying. The conditions of manufacture have a considerable influence on the properties of the dried product and there are many modifications of the standard procedures.

ROLLER-DRYING

Milk to be roller-dried is usually concentrated in a vacuum evaporator until it contains about 17 per cent total solids, and is then homogenized. The liquid is run on to the surface of a slowly-rotating horizontal cylindrical roller which is heated internally by steam under pressure. The temperature of the steam is such that the temperature at the surface of the roller is about 303 °F (150 °C). This causes the water to evaporate rapidly, and after about half a revolution of the roller the milk has formed a thin, dry film which is scaped off by knives mounted against the roller and is then ground to a fine powder. The dried milk is packed in airtight containers.

Roller-drying is a cheaper process than spray-drying, but the product is seldom of as good a quality. It is used mainly to produce skim milk powder for incorporation into animal feeds. Roller-dried milk powders have a slightly 'cooked' flavour. They often do not reconstitute easily and are not completely soluble in water; usually about 85 per cent dissolves. The process largely destroys the structure of the fat globules so that when the powder is reconstituted in warm water much of the fat rises to the surface as an oily layer. However, the relatively severe heat-treatment given to milk during roller-drying ensures a product that is practically free from bacteria. This is probably the main reason why for many years infant foods were based on roller-dried rather than spray-dried milk.

In some modern roller-drying plants the heated roller is enclosed in a vacuum chamber. This enables milk to be dried at the much lower temperature of about 103 °F (40 °C). This is a more expensive process than drying at atmospheric pressure, but the resulting powders are of higher quality and their properties are similar to those of the best spray-dried milks.

SPRAY-DRYING

In this process milk is preheated to 176–194 °F (80–90 °C) for 10–15 seconds, homogenized, concentrated in a vacuum evaporator to about 40 per cent total solids, and then sprayed in the form of a fine mist into the drying chamber where it meets a current of hot, dry air. The temperature of the hot air is around 340 °F (170 °C). The milk particles are dried almost instantaneously and fall to the bottom of the drying chamber as grains of dried milk. The dried milk is run out continuously so that it can be kept cool. It is packed into airtight containers.

The heat-treatment associated with spray-drying is not as severe as in roller-drying. Spray-dried powders are almost completely soluble in water and reconstitute to give milk of good flavour and appearance.

INSTANT MILKS

Difficulty is often experienced in reconstituting even highly soluble milk powder because it tends to float on the surface of the water and resist becoming wetted. This can lead to the formation of lumps in the milk.

Instant milk powders have the desirable property of dissolving in cold water within a few seconds of mixing. They are made by a modification of the spray-drying process in which the powder is re-moistened and then dried again. This causes the particles of dried milk to clump together into larger agglomerates which have a porous structure. When these agglomerates are added to water they act like blotting paper in drawing water into contact with the powder and hasten its passing into solution.

The more elaborate equipment required for preparing instant milk powders makes them expensive products, and it remains to be seen whether these costlier instant milks will capture a substantial part of the market for dried milk. Most instant milk being sold at present is produced from skim milk.

NUTRITIVE VALUE OF DRIED MILK

The composition of good-quality dried whole milk and skim milk are shown in Table 5.1. There is no loss of nutritive value of the major nutrients during properly controlled spray- or roller-drying, and the losses of vitamins are small and similar to those occurring during pasteurization.

TABLE 5.1
Composition of dried milk powder (per 100 g)

	Whole milk	Skim milk
Water (g)	3·0	3·0
Fat (g)	27·5	1·0
Protein (g)	25·0	36·0
Carbohydrate (g)	37·5	50·5
Calcium (g)	0·9	1·3
Vitamin A (retinol equivalents, μg)	240	10
Vitamin D (μg)	0·15	0
Thiamin (μg)	300 †	450 †
Riboflavin (μg)	1350	2000
Folic acid (μg)	40	60
Vitamin B_{12} (μg)	2·5	3·7
Vitamin C (mg)	12 †	18 †
Energy content		
(kJ)	2500	1500
(kcal)	600	360

† Implying a loss of about 15 per cent during processing.

If the milk or milk powder is allowed to become overheated during the drying process, the powder will be yellow or brown in colour, the nutritive value of the proteins will be reduced, and up to one-third of the vitamin B_{12} will be destroyed. The impairment of the proteins results from the so-called 'non-enzymic browning' or Maillard reaction which occurs between lysine in the milk proteins and lactose. The effect is to reduce the amount of lysine that can be released from the proteins during digestion and hence to reduce their nutritive value by rendering part of the lysine unavailable. This can be shown in laboratory tests but it may not be evident from determinations (made using animals) of the biological value of the proteins of overheated milk. This is because milk proteins have an abundance of lysine and do not show themselves to be deficient in this amino acid until about half of it has been rendered unavailable. The loss of available lysine is much more serious when milk proteins are being used as a source of lysine to complement other proteins, such as those of cereals, which are deficient in lysine.

KEEPING QUALITY OF DRIED MILK POWDER

Dried milk powders are hygroscopic, i.e. they readily absorb moisture from the atmosphere. To prevent this, milk powders used to be packed in large tins with soldered tops, but the usual containers nowadays are multi-wall paper sacks lined with high-density polythene, which have proved cheaper and equally satisfactory alternatives for bulk packaging. When the powders are stored at room temperature in such airtight, moisture-proof containers, whole milk powder will keep in good condition for up to a year and skim milk powder for several years. Higher temperatures or absorption of moisture increases the rate of deterioration: the powders develop a stale flavour, become less soluble, and the nutritive value of their proteins fall due to the same non-enzymic browning reaction that occurs when overheating occurs during drying.

Whole milk powders deteriorate more quickly than skim milk powders because oxidative changes in the fat lead to unpleasant tallowy flavours. These changes can be slowed down by displacing, with an inert gas such as nitrogen, all the air from the containers in which the powder is packed. Such gas-packing is essential for whole milk powder destined for use in tropical countries.

6 Infant foods

Breast milk is the natural food for the human infant, but many mothers either do not wish to breast-feed or are unable to produce sufficient milk to satisfy their child. In such circumstances a substitute for breast milk is necessary. In European countries and in the USA, cow's milk forms the basis of nearly all the preparations now used.

COMPOSITION OF HUMAN MILK

During the first few days after birth, human mammary glands, like those of the cow, secrete colostrum—a milk-like fluid much richer in protein than true milk. The colostral proteins contain a large proportion of immune globulins, a component of the whey proteins, that provide the infant with antibodies to combat infections. Unlike the calf, the human infant also receives a good supply of these antibodies from its mother's blood whilst it is in the uterus and before it is born.

The composition of the fluid secreted gradually changes during the first week, and the protein content falls to that typical of human milk, shown in Tables 6.1 and 6.2. The principal differences between the composition of human and cow's milk are that human milk contains much lower contents of protein, calcium, and phosphorus and a much higher content of lactose.

TABLE 6.1
Major constituents of human and cow's milk (per 100 g milk)

	Human	*Cow*
Water (g)	87·5	87·6
Fat (g)	4·6	3·8
Protein (g)	1·2	3·3
casein	0·6	2·6
whey proteins	0·6	0·7
Lactose (g)	6·9	4·7
Calcium (g)	0·03	0·12
Phosphorus (g)	0·015	0·10
Energy content		
(kJ)	310	276
(kcal)	73	66

TABLE 6.2
Vitamins in human and cow's milk (per 100 g milk)

	Human	Cow
Retinol (μg)	46	30
Carotene (μg)	18	20
Vitamin D (μg)	0·01	0·02
Thiamin (μg)	15	40
Riboflavin (μg)	50	180
Nicotinic acid (μg)	170	80
Pantothenic acid (μg)	250	320
Vitamin B_6 (μg)	10	40
Biotin (μg)	1	2·5
Folic acid (μg)	5	6
Vitamin B_{12} (μg)	0·02	0·4
Vitamin C (mg)	4	2

These differences can be related to the much slower rate of growth of the human baby than the young calf: the needs for body-building protein and bone-building calcium are much less in the baby. The higher contents of lactose and fat in human milk give a total energy content similar to that of cow's milk.

The triglycerides of human milk fat differ from those of cow's milk fat in that they contain no short-chain fatty acids but have a greater proportion of long-chain unsaturated fatty acids. Human milk fat is more efficiently digested and absorbed by young babies than are other fats.

The vitamin content of human milk depends on the mother's diet. She can use the reserves of vitamin A in her liver to maintain the vitamin A content of her milk but she is unable to store the water-soluble vitamins and if her diet should be deficient in, for instance, thiamin, riboflavin, or vitamin C, the levels of the vitamins in milk will fall.

There is very little vitamin D and iron in either human or cow's milk.

MILK FOODS FOR INFANTS

Cow's milk can be fed to babies but if the composition of human milk represents that of the best diet for infants, then the differences in composition between human and cow's milk should serve as a guide to how to alter cow's milk to make it a satisfactory food for infants. Various alterations can be made; a wide variety of preparations is now available and new ones are being developed. The choice of which to use is often a personal one. Most babies will remain healthy and grow normally when given any of the preparations made up in the manner recommended by the manufacturer. A minority of babies may have more exacting requirements that will be met more satisfactorily by one diet than by another.

INFANT FOODS

The simplest way of using cow's milk for feeding infants, though not generally recommended for young babies, is to heat pasteurized milk to boiling, cool it, and add one-third by volume of water and 1 teaspoon (5 g) of sugar (sucrose) per $\frac{1}{6}$ pints (100 ml) of diluted milk. Babies fed in this way must be given additional supplies of vitamins C and D.

The most popular preparations for home use are whole- and half-cream milk powders and evaporated whole milk. 'Full-cream' and 'Half-cream' National Dried Milk, as sold in the UK, are roller-dried powders fortified with 4·4 μg of vitamin D per 100 g of powder. Sales of National Dried Milk have declined in recent years; one of the reasons for this is the difficulty of reconstituting the powder without the formation of small lumps which block the teat.

A number of proprietary brands of milk powder are widely used. In several of these the chemical composition of the product has been modified either before or after drying to bring the fat, protein, and carbohydrate contents nearer to those present in human milk. In some brands adjustment is made in a simple manner by altering the fat content or by adding sugar (lactose or sucrose). In other brands, sometimes called 'humanized baby foods', more elaborate alterations are made. These may include some or all of the following: replacement of cow's milk fat by a mixture of animal and vegetable fats chosen to make the fatty-acid composition as similar as possible to that of human milk fat; removal of some of the calcium and phosphorus from cow's milk; and the addition of whey proteins to cow's milk to produce the 1:1 ratio of casein to whey proteins found in human milk. The products may be either roller- or spray-dried and are generally fortified with iron, vitamin A, and vitamin C, as well as with vitamin D. Evaporated milk for infant feeding should contain added vitamin D.

Dried milk powders must be reconstituted and evaporated milk diluted with water before they are given to the baby. Particular care should be taken to follow the instructions on the container and to measure out carefully the recommended amounts of powder or evaporated milk. The baby will not benefit and is likely to be harmed by making its feeds either too concentrated or too dilute. Scrupulous attention to cleanliness is essential. The water should be boiled and allowed to cool in a covered container before use; all containers and utensils used for mixing and the feeding bottle and teat should be absolutely clean and have been sterilized with boiling water. Feeds should be mixed as they are required and they should not be stored. Open tins of evaporated milk and mixed feeds can readily become contaminated by micro-organisms which may multiply rapidly during storage.

Dried milk powders are easy to use in the home and they provide a relatively inexpensive food for infants. They are proving less convenient in maternity hospitals where much work is generated in mixing, dispensing, and sterilizing feeds, and in washing-up the equipment and feeding bottles. There is considerable interest, therefore, in the recent development and production of ready-mixed feeds which are supplied in small disposable bottles to which a teat can be attached. The contents of the bottles are in-bottle sterilized and each bottle contains a single feed for one baby.

7 Cheese

Cheese is prepared from the curd formed when milk is coagulated with lactic acid or rennet, and the liquid (whey) is drained off. Rennet is an extract of calf stomach and it contains the milk-clotting enzyme rennin. This enzyme shares with acid the property of causing the casein in milk to coagulate. The coagulating casein traps the milk fat, so that the curd contains the casein and fat, and the whey contains the lactose, the whey proteins, and much of the mineral content.

Cheese has been used as a food for many centuries and records of its use date back to biblical times. It is quite possible that the first cheese was made by accident when milk was being carried in a pouch made from the stomach of an animal and the enzymes of the stomach converted the liquid milk into a solid mass (or junket)

The manufacture of cheese has long been recognized as a convenient means of converting the fat and protein of milk into a palatable and nutritious product that will keep well. Most cheese is made from cow's milk because cows are more commonly milked than are other animals, but cheeses are also made from ewe's, goat's, and buffalo's milk. In former times cheese was made principally in farm dairies and usually only during periods of maximum milk production in the spring and early summer. Many cheeses are named after towns or geographical regions, and in most cases the name correctly indicates the place of origin. Nowadays most cheese is produced in factories, and several stages of the cheesemaking process have been mechanized.

The consumption of cheese in the UK has been steadily increasing and each year each of us now eats about 15 lb (7 kg). At present, about half of this cheese is made in the UK (90 per cent in factories, 10 per cent farmhouse), and the remainder is imported.

More than 400 varieties of cheese are known, though only a small proportion of these are familiar to us in the UK. Each variety is made by a unique process. The differences between varieties in appearance, flavour, texture, and composition depend on the precise way in which the curd is formed, treated and ripened by the action of bacteria or moulds.

Most of the cheese sold in the UK falls into one of three main types: hard and semi-hard, blue-veined, and soft. In addition there are cottage, cream, and whey cheeses.

HARD AND SEMI-HARD CHEESES

Cheddar

Cheddar is the best-known hard cheese. More than half the cheese made in the UK is of this type and further supplies are imported so that in total it amounts for about two-thirds of the cheese sold.

Cheddar cheese is usually made from pasteurized milk. The first step is the addition of a 'starter'. This is a culture of pure strains of two lactic-acid-producing bacteria, *Streptococcus lactis* and *Streptococcus cremoris*. These bacteria ferment milk lactose and produce lactic acid. As soon as sufficient acidity has been produced the acidified milk is warmed and rennet is added. A semi-solid curd forms within a few minutes. This is cut by knives into small pieces which are heated slowly to about 104 °F (40 °C) and stirred. This causes the pieces of curd to shrink and become more granular and to express the whey. The whey is then run off and the curd forms a dense rubbery mat which is cut into blocks. These are piled up and allowed to stand for about 2 hours to allow the curd to become firmer and more whey to drain off. This is the 'cheddaring' process; it produces a close-textured cheese compared with the 'crumbly' texture of Cheshire and Caerphilly. The curd is then milled, i.e. it is broken into small pieces, which are mixed with salt (about 3·5 oz (100 g) to 11 lb (5 kg) curd). The salted curd is filled into moulds and pressed overnight. Traditionally, Cheddar cheese is pressed in cylindrical moulds, each holding about 70 lb (32 kg) of cheese, but rectangular moulds holding about 40 lb (18 kg) of cheese are now being used to an increasing extent. Finally the pressed cheese is bandaged with a cotton cloth or wrapped in plastic film and stored at 49 °F (10 °C) for 3 months to allow it to ripen. Ripening is brought about by bacterial enzymes which convert any residual lactose to lactic acid and break down some of the fat and protein of the cheese to form the various compounds that give the mature cheese its characteristic flavour. A mature Cheddar cheese should remain in good condition for at least a year.

The yield of cheese obtained depends on the fat and casein content of the milk used, but it generally requires about 1 gallon (4·54 l) of milk to produce 1 lb (0·45 kg) of cheese.

Other varieties

Other varieties of hard and semi-hard cheese are made by methods which can be regarded as modifications of the procedure used for making Cheddar cheese. For each variety the cheesemaker has selected treatment conditions that will give a curd having the acidity and moisture content which will allow the development during ripening of the required flavour and texture. Almost every stage of the cheesemaking process can be modified but most of the different characteristics of cheese varieties result from differences in

CHEESE

the fineness of cutting the curd during milling, in the temperature to which the curd is heated, and in the pressure applied during pressing.

Readily available British and imported varieties include the following.

British.

Caerphilly: a white cheese that matures about 2 weeks after manufacture. It is only lightly pressed and has a moist, springy texture and a mild flavour. It deteriorates rapidly after about 4 weeks due to its high moisture content.

Cheshire: usually a white cheese, but it is sometimes coloured red by adding annatto, a vegetable dye obtained from the fruit of the annatto tree. It is milder in flavour and has a more crumbly and softer texture than Cheddar. Cheshire cheeses ripen in 4 weeks and should remain in good condition for up to 6 months.

Derby: a white, hard cheese with a smooth texture and a soft, mild flavour. Derby cheeses are made in a flat circular shape 14 in (36 cm) in diameter and 5 in (13 cm) high; they ripen in about 4 months and remain in good condition for up to 6 months.

Double Gloucester: a cheese that closely resembles Cheddar but has a slightly milder flavour. Double Gloucester cheeses are 14 in (36 cm) in diameter and about 8 in (20 cm) high, they ripen in 4–6 months; the mature cheese should remain in good condition for at least a year. In earlier days a smaller, quicker-ripening cheese called Single Gloucester was also made.

Lancashire: a semi-hard cheese with a soft texture and mild flavour. It spreads easily and is popular for toasted cheese and Welsh Rarebit. The method of manufacture differs from that for most other cheeses in that part of the curd from one day's making is held over and mixed with that from the following day's milk. The cheeses ripen in about 2 months and should be eaten within the following 6 months.

Leicester: a bright orange-red cheese (coloured with annatto) that has a loose, flaky texture and rich flavour. The cheeses ripen in about 2 months and remain in good condition for up to a year.

White Wensleydale: a semi-hard cheese having a soft, crumbly texture and a mild flavour. The cheeses are 8 in (20 cm) in diameter and 6 in (15 cm) high, they ripen in about 2 weeks and remain in good condition for several months.

Imported.

Emmenthal and Gruyère: rather similar hard cheeses with mild, sweetish flavours. Originally made in Switzerland but now in many countries.

Characterized by large number of holes or eyes which are formed as reservoirs for the carbon dioxide produced by the action of Propionibacteria during ripening, which takes about 6 months. These cheeses remain in good condition for at least a year.

Edam: a semi-hard cheese of Dutch origin with a soft and moist texture and not much flavour. Usually made from partly skimmed milk and therefore low in fat. Often shaped like small flattened footballs with a bright red wax coat.

Gouda: another Dutch cheese; resembles Edam but made from whole milk and so of higher fat content.

Parmesan: a very hard, long-keeping cheese made in Italy from partly skimmed milk; mainly used for grating and cooking.

BLUE-VEINED CHEESES

These are varieties made from a soft curd which is broken into small pieces which are packed gently into moulds but not pressed. The unripe cheeses are kept in a damp, cool store and a blue mould, *Penicillium roquefortii*, is allowed to grow through cracks in the cheese or, more usually, the mould is introduced by stabbing the cheese with a long needle. This mould assists in ripening and imparts a particular flavour to the cheese; ripening should be slow and requires up to 6 months.

The best-known varieties are the English-made Stilton and Blue Wensleydale, Danish Blue from Denmark, Gorgonzola from Italy, and Roquefort (a ewe's milk cheese) from France.

SOFT CHEESE

In making these cheeses the coagulated curd is not cut into pieces but is ladled intact into shallow round forms and allowed to drain. When the curd has dried out it is rubbed with salt and placed for several weeks in a dark store where it is allowed to become covered with a whitish growth of the mould *Penicillium candidum*. Enzymes from this mould diffuse into the body of the cheese and act on the curd to produce the smooth soft paste and characteristic flavour of the ripened cheese.

Most cheeses of this type originated in France and two of the best-known are Brie and Camembert. The interiors of properly made Brie and Camembert cheeses should become uniformly soft as the cheese ripens fully. Once ripe the cheeses should be eaten within a few days.

Although the rinds of most cheeses are not edible, those of soft cheeses can be pleasant to eat; they are also a good source of B vitamins (see p. 39).

Cottage cheese

This is an unripened soft cheese made by coagulating skim milk with acid. It has a soft, smooth texture and a bland flavour, and it has become a popular product. Cottage cheese is prepared by inoculating pasteurized skim milk with cultures of the lactic acid-producing bacteria, *Steptococcus lactis* and *Streptococcus cremoris,* and then cutting, heating, washing, and draining the resulting curd. Cottage cheese is usually packed in small cartons. It should be stored in a refrigerator and used within a week or so of manufacture. It does not ripen or improve during keeping because its low acidity and its moist, open texture allow spoilage micro-organisms, particularly moulds, to grow and multiply.

Cream cheese

Cream cheese is made from single or double cream by adding cultures of lactic-acid producing bacteria and rennet. The resulting coagulum is drained through a fine cloth, lightly pressed, and then packed into cartons. The cheeses are soft and smooth in texture and contain very little besides fat and water. The single-cream variety contains about 36 per cent and the double cream cheese about 72 per cent fat.

Whey cheese

This is made in several European countries, but not in Britain, by precipitating the whey proteins by heating acid whey. The Italian Ricotta and the Scandinavian Mysost and Gjetost are three of many different varieties of whey cheese.

Mysost is made from cow's milk whey and Gjetost from a mixture of cow's and goat's milk whey. The whey is concentrated by boiling to about a quarter of its original volume. The mass is stirred whilst cooking and set in small moulds. The products contain only about 10 per cent of water and consist mainly of caramellized lactose (up to 40 per cent) together with the other constituents of whey; they keep almost indefinitely.

PROCESSED CHEESE

This is not a variety of cheese but a manufactured product made usually from ripened hard or semi-hard cheese of inferior quality. The cheese is emulsified with salts, usually citrates and phosphates, and heated to about 185 °F (85 °C). Other substances such as dried skim milk powder or whey powder may be added and the mixture may be flavoured or coloured. The heating process kills most of the bacteria present, so the product, which is usually wrapped in metal foil or plastic film, remains moist and keeps well.

Similar, but softer preparations containing more moisture are also made and are often called cheese spreads.

FILLED OR VEGETABLE-FAT CHEESES

Cheese can be made by the normal methods from milk in which butterfat has been replaced by cheaper vegetable fats. Such substitution should produce a cheaper cheese; it can also provide a cheese containing a higher content of polyunsaturated fatty acids if a vegetable fat such as maize oil or sunflower oil is used (see p. 46).

CARE OF CHEESE IN THE HOME

Cheese is at its best when eaten freshly cut from the whole cheese but portions of cheese will retain their flavour and moisture for a week or two if wrapped in polythene bags and stored in a refrigerator or cool larder. Cheese stored in a refrigerator will tend to be tasteless unless it is taken out an hour before serving to allow it to warm up to room temperature.

THE COMPOSITION OF CHEESE

The curd formed in the first stages of cheesemaking is composed of about half the total solids of whole milk. It contains virtually all the fat and casein, about two-thirds of the calcium, most of the vitamin A, one-quarter of the riboflavin and about one-sixth of the thiamin in the original milk. The lactose, whey proteins, and the rest of the minerals and vitamins remain in the whey (Table 7.1). The curd formed during the making of skim milk cheese contains only about one-third of the solids of the original milk; the losses of lactose, minerals and vitamins in the whey are very much the same as those during cheesemaking from whole milk.

TABLE 7.1
Percentages of nutrients in curd and whey from whole milk

	Curd	*Whey*
Fat	94	6
Protein	75	25
casein	96	4
whey	4	96
Lactose	6	94
Calcium	62	38
Vitamin A	94	6
Thiamin	15	85
Riboflavin	26	74
Folic acid	5	95
Vitamin B_{12}	25	75
Vitamin C†	6	84
Total solids	48	52

† Some vitamin C is destroyed by the action of light.

The fate of the nutrients in the curd during ripening depends on the type of cheese. The compositions of ripened cheese of different types are shown in Table 7.2. Hard and semi-hard cheeses may lose small amounts of nutrients in the whey expelled during pressing, but the activity of bacteria during ripening leads to the synthesis of several water-soluble vitamins. Blue-veined cheeses receive an additional contribution to their content of water-soluble vitamins from the growth of the *Penicillium* mould. Much of the vitamin content synthesized during the ripening of soft cheeses remains in the rind and may contribute little nutritionally since the rind is often not eaten.

TABLE 7.2
Composition of cheese of various types (per 100 g)

	Hard Cheddar	Semi-hard Edam	Blue-veined Roquefort	Soft Camembert	Cottage cheese
Water (g)	35	43	40	51	79
Fat (g)	33	24	31	23	0·4
Protein (g)	26	26	21	19	16·9
Calcium (g)	0·83	0·76	0·32	0·38	0·09
Vitamin A (retinol equivalents, µg)	380	250	300	240	3
Thiamin (µg)	50	60	30	50	30
Riboflavin (mg)	0·50	0·35	0·70	0·45	0·28
Energy content (kJ)	1670	1330	1500	1180	340
(kcal)	400	320	360	280	82

Calcium content of various cheeses
Hard and semi-hard cheeses contain more calcium than the others because there is little loss of the mineral after the inital separation of the curd and whey. But in the preparation of blue-veined, soft, and cottage cheeses, more than half the calcium initially present in the curd is lost when the whey is allowed to become more acid and to drain away slowly, so their calcium content is much lower (see Tables 7.1 and 7.2).

Value of cheese in the diet
Cheese is widely recognized as an excellent food. It can be eaten alone, with bread or biscuits, or it can be incorporated into a wide range of cooked dishes.

All cheeses contain a high proportion of protein (see Table 7.2) and all except cottage cheese are good sources of calcium and are rich in fat, a good source of energy. Cheese is a concentrated food and, weight for weight, the energy content of hard cheese is the same as that of sugar.

The cheese protein, casein, is slightly inferior to the proteins of whole milk (see p. 10), but it is an excellent source of lysine which enables it to complement the proteins of cereals such as maize or wheat. This was clearly shown in an experiment in which rats were given bread and cheese and the biological values of the proteins in each were determined. The results (Table 7.3) confirmed that bread protein was inferior to cheese protein, but when bread and cheese were given together the value of the combined protein was as high as that of cheese. A further point of particular interest was that when the rats were given bread and cheese on alternate days there was very much less improvement in the biological value of the proteins. Similar experiments have been carried out with human subjects and it is now well established that proteins effectively complement each other only when they are eaten at the same meal or with a relatively short interval between eating them. It is sensible, therefore, to eat bread with your cheese.

TABLE 7.3
The biological value for rats of the proteins of bread and cheese

	BV
Bread	53
Cheese	76
Bread and cheese (given together)	76
Bread and cheese (given on alternate days)	67

Cottage cheese has become increasingly popular as a 'low calorie, protein-rich' food. This is a true description of the product. Its energy content is only a fifth of that of hard cheese, though unfortunately its calcium content also is much lower than that of other types of cheese.

WHEY

Whey is the liquid remaining after the separation of the curd in cheesemaking. It contains the whey proteins, lactose, and much of the minerals from the original milk. The composition of Cheddar cheese whey is shown in Table 7.4. Wheys from other cheese are similar in composition, but those from acid curds contain more calcium. The small amount of fat in whey is usually separated off and made into whey butter.

The 300 million gallons (1400 million litres) of whey produced each year in the UK contain about 12 000 tons (12 million kg) of high-quality protein and about 60 000 tons (60 million kg) of lactose. These nutrients, particularly the protein, are valuable as human food and one of the major problems of the dairy industry is to find efficient and economic methods of processing this vast quantity of whey. Its disposal by discharge into rivers is wasteful and causes pollution.

At present much of the whey produced is used for animal feeding and either sold at a very cheap price to pig farmers who use it as a liquid feed, or spray- or roller-dried and used as an ingredient in diets for young calves.

For human nutrition, dried whey is used in the baking industry, in the manufacture of whey soups and sauces and of fruit whips. An increasing industrial use of whey is as a source of lactose which is used by the pharmaceutical industry to dilute drugs in pills.

TABLE 7.4
Composition of Cheddar cheese whey (per 100 g)

Water (g)	93·3
Fat (g)	0·3
Protein (g)	0·9
Lactose (g)	4·7
Calcium (g)	0·05
Retinol (μg)	2
Thiamin (μg)	40
Riboflavin (μg)	80
Folic acid (μg)	5
Vitamin B_{12} (μg)	0·15
Vitamin C (mg)	1

8 Cream, butter, and margarine

CREAM

Cream is not a specifically definable substance, being simply milk in which the content of fat has been greatly increased. Cream can be obtained by skimming off the layer that rises to the surface of fresh milk when it is allowed to stand. It is more usually prepared by centrifuging milk in a cream separator. Cream is less dense than milk, so when milk is run into the rapidly spinning bowl of the separator the skim milk goes to the outside of the bowl and the cream remains in the centre. An arrangement of plates in the spinning bowl allows the skim milk and cream to be removed continuously. The fat content of the cream can be regulated by adjusting the rate at which milk is run into the bowl.

Three types of fresh cream are produced: single cream, containing 18 per cent fat; double cream, with 48 per cent fat; and whipping cream, with 35 per cent fat. All types are pasteurized by the HTST process at 180–190 °F (82–88 °C) for 10 seconds, cooled to 40 °F (5 °C) and filled into cartons. Single cream is usually also homogenized. Cartons of cream kept unopened in a refrigerator will remain in good condition for a week or more.

Sterilized cream is prepared from cream containing 23–27 per cent fat which is homogenized, filled into bottles or cans, and heated to about 240 °F (115 °C) for 20–30 minutes. Sterilized cream in unopened containers will keep for some months though, as with evaporated milk, some thickening of the product occurs during prolonged storage.

The whipping of cream

Cream containing between about 30 per cent and 38 per cent fat can be whipped to produce a stable foam by trapping tiny bubbles of air in the cream. At the same time the volume of the cream is increased two- or three-fold. For best results both the bowl and the cream should be cooled to below 45 °F (8 °C).

A good substitute for whipping cream can be made by diluting 3 parts of double cream with 1 part of milk. Since homogenizing destroys the whipping property of cream, a mixture of homogenized single cream and double cream does not whip satisfactorily.

Clotted cream

Clotted cream is also called Devonshire or Cornish cream. To make it milk is poured into shallow pans and left to stand for 10–12 hours to allow the

CREAM, BUTTER, AND MARGARINE

cream to rise. The pans are then heated to raise the temperature of the milk and cream to about 180 °F (82 °C). After this heating or scalding the pans are allowed to cool slowly, and about 24 hours later the blanket of cream is skimmed off and packed into cartons.

A good-quality clotted cream should have a characteristic scalded flavour, be granular in texture, and contain 62–65 per cent of fat.

Home-made cream

The present day price structure of dairy products results in butter being a much cheaper source of milk fat than cream. Advantage can be taken of this to make a perfectly satisfactory and very acceptable cream containing about 40 per cent of fat from unsalted butter and milk. This can be done at home by melting 3½ oz (100 g) of unsalted butter and half a small level teaspoonful (2 g) of powdered gelatin in $\frac{1}{6}$ pint (100 ml) of warm (100 °F, 38 °C) milk (preferably Channel Island). The mixture is mixed in a liquidizer (previously warmed with warm water) at top speed for 1½ minutes, then poured into a jug, covered, and stood in a refrigerator overnight. The result is a product ($\frac{1}{3}$ pint, 200 ml) that is not easily distinguished from double cream.

Larger quantities of cream can be made using similar proportions of ingredients and passing the warmed mixture through a piston-type homogenizing machine.

The nutritive value of cream

The composition of different types of cream is shown in Table 8.1. Cream consists of an emulsion of milk fat in skim milk, and its nutritive value is that of these two components. Since fat is the main nutrient, cream has a high content of energy: 1 oz (30 g) of single, or ½ oz (12 g) of double cream contains as much energy as 3¾ oz (100 g) of milk.

TABLE 8.1
Composition of different types of cream (per 100 g)

	Single	Double	Whipping	Clotted
Fat (g)	18	48	36	63
Protein (g)	3·0	1·6	2·2	3·0
Lactose (g)	4·1	2·4	3·1	2·5
Water (g)	74	48	58	30
Energy content				
(kJ)	790	1900	1400	2450
(kcal)	190	450	335	590

BUTTER

Butter is made by churning cream. The churning action causes the fat-in-water emulsion of cream to break down and change into the water-in-fat emulsion of butter. The traditional wooden churns have now been replaced in many countries by an automatic continuous churning process.

Fresh cream is used for buttermaking in most countries, including the UK and New Zealand, but in Denmark the cream is ripened or soured by lactic-acid-producing bacteria before use. Most butter contains 1·5—2 per cent of added salt, though unsalted butter is also made and is preferred by some people. Unsalted butter is often more suitable than salted for use in cooking, particularly in the preparation of butter sauces.

Cream used for buttermaking contains about 34 per cent fat. It is pasteurized at a high temperature (203 °F, 95 °C), cooled, and run into the churn which is rotated. After about 30 minutes the butter separates in the form of small grains which are strained off from the liquid, called buttermilk, in which they are suspended. The butter grains are washed with water and drained. Salt is added if salted butter is being made, and the grains are then kneaded or 'worked'. This process causes them to coalesce into a compact mass of butter. Properly worked butter has a firm, waxy body which is not greasy and contains no visible droplets of water. Inadequate control of the manufacturing procedure, particularly of the churning temperature and working, can easily result in inferior quality butter having a greasy, gritty, or crumbly texture.

For storage in bulk, butter is usually packed into parchment-lined boxes that hold 56 lb (25 kg). Butter so packed can be stored at −10 °F to −20 °F (−25 °C to −30 °C) for a year or more without change in flavour. For retail sale, butter is usually cut into ½ lb (227 g) blocks which are wrapped in foil-lined-parchment or greaseproof paper.

Foil-lined-parchment wrappings is to be preferred since both in the shop and in the home butter should be protected from light, which causes it to go rancid and which destroys vitamin A, and also from the opportunity of picking up odours or flavours from other foods.

Salted butter keeps better than does unsalted, and butter made from fresh cream keeps better than that made from ripened cream.

The nutritive value of butter

The composition of butter varies little and typical values for salted and unsalted butters are shown in Table 8.2. Butter sold in the UK must contain not less than 80 per cent fat and not more than 16 per cent of water.

The nutritional value of butter depends almost entirely on its content of fat and fat-soluble vitamins, particularly carotene and retinol. The natural pale yellow colour of butter is due to its content of carotene, so that butter made from summer milk will be yellower than that made during the winter. The protein, lactose, minerals, and water-soluble vitamins of milk are almost entirely absent from butter.

Butter, like cream, is an energy-producing food and yields about 730 kcals per 100 g (3000 kJ per 100 g). It is also an excellent source of vitamin A; ½ oz (14 g) of butter supplies a child with about one-third of his daily requirement of this vitamin.

TABLE 8.2
Composition of butter and margarine (per 100 g)

	Butter		Margarine
	Salted	Unsalted	
Fat (g)	81	83	81
Water (g)	16	16	16
Salt (g)	2·0	–	2·0
Milk solids (protein, lactose, minerals) (g)	0·9	1·0	0·9
Vitamin A (retinol equivalents, μg)	700	720	700
Vitamin D (μg)	0·5	0·5	8
Energy content			
(kJ)	3000	3100	3000
(kcal)	730	750	730

MARGARINE

Margarine has an obvious and close resemblance to butter, both in composition (Table 8.2) and in properties. It is made by emulsifying skim milk into a mixture of purified oils and fats from plants, fish, or other animals, with the addition of salt, flavouring, colouring, and vitamins A and D. Some margarines may also have up to 10 per cent of butter added.

The particular oils and fats used by the margarine manufacturer depend on the price and availability of supplies. Those most commonly used at present in the UK include fish oil (usually herring), soya-bean oil, palm oil, coconut oil, rape-seed oil, and sunflower-seed oil.

Oils differ from fats only in that they have a lower melting point and are liquid at room temperature. Oils and fats are composed of mixtures of triglycerides (see p. 8). The triglycerides of oils usually contain a high proportion of unsaturated fatty acids and have lower melting points than the triglycerides of fats in which saturated fatty acids predominate. Before fish and plant oils are used in the manufacture of margarine they are 'hardened', i.e. they are converted into fats by hydrogenation. This is a chemical process by which hydrogen is added to some of the unsaturated fatty acids in the presence of a finely powdered metal (usually nickel) catalyst. Hydrogenation converts a proportion of the unsaturated fatty acids into saturated fatty acids and thus raises the melting point of the triglycerides.

Butter and margarine each contain about 80 per cent fat and the principal difference between them is in the amounts and composition of the various triglycerides making up this fat. The component triglycerides of milk, and hence of butter, vary only slightly, so that wherever and whenever butter is made it contains essentially the same mixture of triglycerides, and hence always melts over the temperature range 82–100 °F (28–38 °C). The manufacturer of margarine can adjust the melting point of his product by selecting and processing a mixture of oils, hydrogenated (hardened) oils, and fats to obtain either a soft margarine ('spreads straight from the fridge') with a lower melting point than butter, or a hard margarine with a similar melting point to that of butter.

Margarine, like butter, should be stored in a cold place away from light and in a sealed wrap or container to prevent it picking up odours or flavours from other foods. Hard margarines are usually sold in blocks and soft margarines are packed in tubs. Hard margarines are largely used by bakers for making pastry.

The nutritive value of margarine

Margarine is a wholesome food with essentially the same nutritive properties as butter, to which it is in no way inferior.

Various flavouring agents, including diacetyl—the main contributor to the flavour of butter—are added to margarine to make its flavour as similar as possible to that of butter. Nonetheless, most people can distinguish the flavour of butter from that of margarine and prefer the flavour of butter.

Recent research work has shown that diets in which most of the fats contain mainly saturated fatty acids may contribute to the accumulation of undesirably high levels of cholesterol in the blood of some people, and that this may in turn lead to the development of diseases of the heart and blood vessels, such as atherosclerosis. However, blood cholesterol levels are lowered when diets are eaten which contain fats with a high proportion of polyunsaturated fatty acids, that is fatty acids, such as linoleic acid (see p. 8), with two or more pairs of hydrogen atoms missing. Butter has a low content (about 2 per cent) of polyunsaturated fatty acids and so have many hard margarines (5–10 per cent), but certain special margarines are available that contain high contents (40 per cent) of polyunsaturated fatty acids. The special margarines include a high proportion of vegetable oils such as sunflower-seed oil or safflower-seed oil which are rich in polyunsaturated fatty acids.

9 Ice cream

The name 'ice cream' is often used rather loosely to cover a wide variety of frozen desserts, ranging from the true ice creams that can definitely be classed as foods to the sorbets, water ices, and iced lollies that are simply refreshing sweets made from iced fruit juices but that are of little nutritive value.

In the strict sense, ice cream is a milk product and is made from cream (or vegetable fat), milk, sugar, and flavouring. It is required to contain not less than 7·5 per cent of milk solids other than fat and not less than 5 per cent of fat. In products called dairy ice cream, dairy cream ice, or cream ice, the whole of fat must be milk fat.

An infinite variety of different types and flavours of ice cream can be made and in the UK the industry that has grown up around this food produces each year about 50 million gallons (230 million litres) of ice cream, 12 million gallons (55 million litres) of dairy ice cream, and 14 million gallons (63 million litres) of water ices.

Commercially made ice creams can be divided into two main categories: hard ice cream and soft ice cream. Hard ice creams are mainly produced by the larger manufacturers and distributed through retail shops where they are sold after storage in deep-freeze cabinets at $-4\,°F$ to $-20\,°F$ ($-20\,°C$ to $-30\,°C$). Soft ice creams are also made by the large manufacturers but they are often made each day in small batches by local traders for immediate sale from shops or mobile vans; soft ices are only lightly frozen and are stored at around $24\,°F$ ($-5\,°C$).

The same basic ingredients (Table 9.1) are used in the preparation of both hard and soft ice creams, though the proportions of the different ingredients in the mixes differ, for instance hard ice creams usually contain relatively more fat and less milk solids than soft ice creams.

HARD ICE CREAM

The ingredients that could be used to prepare a typical good-quality hard ice cream are shown in Table 9.2. The solid ingredients are dissolved or dispersed in warm water and the mixture is pasteurized or UHT-sterilized by procedures similar to those used for milk but at slightly higher temperatures. The mixture is then homogenized, cooled, and frozen at about $22\,°F$ ($-5\,°C$) either in a batch or continuous freezer. During freezing the mix is vigorously stirred to incorporate air into it. Air is a necessary ingredient of ice cream, because without it the mix would freeze to a hard or soggy mass. The percentage increase in volume of the mix effected by incorporating air is known as 'overrun'. For hard ice cream the overrun is usually 90–100 per cent i.e. 1 part of mix makes almost 2 parts of ice cream.

TABLE 9.1
Basic ingredients used in the manufacture of ice cream

	Range of content (per cent)
Fat	6–12
Butterfat	
Vegetable fats	
Milk solids	10–12
spray-dried milk powder	
condensed milk	
Sugar	11–15
sucrose	
cornflour	
glucose	
Stabilizer	
Emulsifier	1
Flavouring	
Colouring	

TABLE 9.2
Percentage composition of hard ice-cream mix

Butter or vegetable fat	11·9
Dried milk powder	10·5
Sucrose	14·25
Emulsifier and stabilizer	1
Water	62·3

The emulsifier helps to disperse the fat globules throughout the mix and the stabilizer improves the 'body' of the product.

Once frozen, the ice cream is cut up into blocks which are packed into suitable containers or wrapped before being hardened in a refrigerator at about −40 °F (−40 °C). Provided that the hardened ice cream is kept at a very low temperature (around −20 °F (−30 °C)) it will remain in good condition for at least six months.

SOFT ICE CREAM

The ingredients that could be used to prepare a soft ice cream are shown in Table 9.3. The ingredients are mixed, pasteurized, homogenized, and frozen at 22 °F (−5 °C) as for hard ice cream, except that less air is incorporated

ICE CREAM

into the mix during freezing. The overrun usually being about 50 per cent, that is 1 part of mix makes about 1·5 parts of ice cream.

TABLE 9.3
Percentage composition of soft ice-cream mix

Vegetable fat	6
Dried milk powder	11·5
Sucrose	13
Emulsifier and stabilizer	1
Water	68·5

An increasing amount of soft ice cream is made by reconstituting suitable pre-mixed ingredients which have been heat-sterilized and which are available as a concentrate that needs only the addition of clean water to give a mix suitable for freezing.

Soft ice cream is usually sold by the scoopful in cornets or cups. It is stored in bulk at a temperature just below 32 °F (0 °C) and is unlikely to remain in good condition for more than a few days.

SHERBETS

A sherbet is a frozen sweet made from a mix containing water, sugar, fruit juice or flavour and colour, and about 2 per cent of milk solids. The frozen product has an overrun of about 40 per cent.

NUTRITIVE VALUE OF ICE CREAM

Ice cream is a palatable and readily digested food though it contains predominantly the energy-supplying nutrients, fat and carbohydrate. Its protein content, typically about 4 per cent, is too low for ice cream to be considered a balanced food. Dairy ice cream provides the carotene, retinol, and vitamin D associated with the milk fat, but other ice creams do not contain the fat-soluble vitamins. No ice creams contain nutritionally significant amounts of the water-soluble vitamins.

Ice cream is normally served by volume rather than by weight and it is of interest that despite differences in composition a typical serving of about one-tenth of a pint (50 ml) of almost any of the various types of ice cream supplies about 200 kJ (50 kcal). This is illustrated in Table 9.4 in which is set out the nutrient content of the ice creams in Table 9.2 and 9.3. When allowance is made for the differences in overrun during freezing, it is evident that equal volumes of the two ice creams supply the same amount of dietary energy.

TABLE 9.4
Nutrient content of ice cream (per cent)

		Hard ice cream	Soft ice cream
Protein		3·8	4·1
Fat		10	6
Carbohydrate		19·75	19
Energy content of total mix (per 100 g)	(kJ)	770	610
	(kcal)	184	146

		Frozen at 90 per cent overrun	Frozen at 50 per cent overrun
Volume of ice cream from 100 g total mix (ml)		190	150
Energy content of ice cream (per 100 ml)	(kJ)	405	405
	(kcal)	97	97

10 Soured and fermented milks

Fresh milk turns sour on keeping. This property has been used by man since the beginning of recorded history to save milk from spoilage by harmful bacteria. The milk is deliberately soured by adding harmless bacteria which produce acid and so control the growth of other bacteria which might be harmful. By introducing particular micro-organisms and controlling their growth and activity by regulating the temperature of the milk, a variety of wholesome, palatable forms of soured or fermented milks can be prepared. Such milks are often called cultured milks and many of them originate in Northern and Eastern European countries where they are still widely made and consumed. Well-known examples include yoghurt, which is the traditional form of sour milk in Bulgaria and neighbouring countries, and kefir and kumiss (the latter made from mare's milk) which are sour, fermented milk products of the Caucasus and the southern steppes of the USSR.

Soured milks are produced by the action of acid-forming bacteria which convert lactose to lactic acid; the resulting acidity causes the milk proteins to clot. For the preparation of sour, fermented milks the action of bacteria is augmented by yeasts which ferment lactose to alcohol (kefir may contain up to 1 per cent alcohol, and kumiss up to 3 per cent of alcohol).

During recent years manufacturing processes have been developed for the preparation of cultured milk products on a large scale and under carefully controlled conditions. Two of these products, yoghurt and cultured buttermilk, have proved popular in the UK.

YOGHURT

Yoghurt is milk which has acquired a characteristic flavour due to the growth of two micro-organisms, *Streptococcus thermophilus* and *Lactobacillus bulgaricus.* Yoghurt is acidic and has a fine smooth texture which can be a firm gel like junket or a viscous liquid like custard, depending on the way in which it is made.

There are, as yet, no compositional standards for yoghurt, and a wide variety of products are made by a number of different manufacturers. The basic ingredients may be whole milk, partially skimmed milk, skimmed milk, evaporated milk, dried milk, or a mixture of any of these products. The mixture selected usually contains rather less fat and rather more non-fatty milk solids than does milk.

To prepare yoghurt the ingredients are warmed and homogenized. The mix is then treated to kill micro-organisms by heating to about 194 °F (90 °C) for 30 minutes, cooled to a temperature of about 110 °F (44 °C) and inoculated with cultures of *Streptococcus thermophilus* and *Lactobacillus bulgaricus*

When a firm yoghurt is to be made the inoculated mix is incubated at 110 °F (44 °C) for 1–1½ hours and then poured into bottles or cartons which are kept in a warm room until the milk has coagulated. The room is then cooled to 50 °F (10 °C) and when the yoghurt has set firm it is transferred to cold storage at a temperature of 40–45 °F (5–8 °C). Stored in this way the product should remain in good condition for 7–10 days.

When a liquid type of yoghurt is to be made, the inoculated mix is incubated at a slightly lower temperature than for firm yoghurt and is constantly stirred so that it thickens but does not form a firm curd. It is then poured into bottles or cartons and cooled and stored in the same way as firm yoghurt.

Natural yoghurts are usually of the firm type and are made simply from milk products, though sugar may sometimes be added to give a sweetened yoghurt. Fruit juices or flavouring essences and sugar are often added to the mix used for the preparation of liquid yoghurts.

Home-made yoghurts

Yoghurt can readily be made on a small scale from 1 pint of sterilized or UHT milk (or boiled and cooled fresh milk), by adding 1¾ oz (50 g) of dried skim milk powder and 1 level tablespoon of natural yoghurt, whisking in a jug or mixing in a liquidizer, and pouring into jars or cartons which are stood in a saucepan of hot water (2 cups boiling and 1 cup cold water), covered with lid, and left until set. When the yoghurt is set firm the jars or cartons should be transferred to and stored in a refrigerator.

The same method can be used with diluted evaporated milk; a richer product can be made by substituting dried whole milk for dried skimmed milk, or by adding cream.

It is important that before use all containers and utensils should be scrupulously cleaned by scalding or washing with detergent and rinsing well.

Nutritive value of yoghurt

The reputed value of yoghurt in ensuring long life and good health was attributed by Metchnikoff to the beneficial activities of its micro-organisms in the intestines of those who ate it. This theory is no longer generally accepted and the nutritional value of yoghurt is considered to be that of the milk used in its preparation.

The composition of milk is compared with that of a typical firm yoghurt and a representative sweetened fruit-flavoured yoghurt in Table 10.1.

TABLE 10.1
The nutrient content of milk, natural yoghurt, and fruit-flavoured sweetened yoghurt (per 100 g)

	Milk	Natural yoghurt	Fruit-flavoured sweetened yoghurt
Fat (g)	3·7	1·5	1·5
Protein (g)	3·2	5·0	4·3
Carbohydrate (g)	4·7	7·1	14·0
Vitamin A (retinol equivalents, μg)	30	12	12
Thiamin (μg)	45	65	55
Riboflavin (μg)	180	270	240
Energy content			
(kJ)	272	259	364
(kcal)	65	62	87

Yoghurts contain more protein, thiamin, and riboflavin than milk but less vitamin A. There is little difference between milk and natural yoghurt in the content of energy-supplying nutrients but, because of the added sugar, sweetened yoghurt is a substantially richer source of energy than milk.

CULTURED BUTTERMILK

Cultured buttermilk is produced by adding a culture of acid-producing *streptococci* to skim milk. It has a mild acid taste and a slightly butter-like flavour. It is similar in properties to the buttermilk produced during the churning of soured cream into butter, and its nutritional value is the same as that of the skim milk from which it is made.

11 Legislation controlling milk and dairy foods

Parliament has passed a considerable number of Acts dealing with the production, processing, and marketing of food. These Food Laws are intended to protect the consumer by ensuring that he is supplied only with products that conform to certain specified standards of compositional and hygienic quality.

Milk has attracted particular attention from legislators for two reasons. First, because it is open to fraudulent sale; in the last century, in the early days of milk sales, it was not unknown for milk to be diluted by the addition of water during its journey from the cow to the consumer. Secondly, because it has become evident that to ensure that milk is free from pathogenic micro-organisms and is of good keeping quality it is necessary to produce it under hygienic conditions and to control microbial contamination by using some form of heat-treatment carried out under prescribed conditions.

The production and marketing of milk and milk products in the UK are now comprehensively controlled by the provisions of the Food and Drugs Act 1955 and the various Regulations that have been issued as Statutory Instruments under this Act. Separate but similar Regulations are issued for England and Wales, for Scotland, and for Northern Ireland. The Regulations are printed and published by H. M. Stationery Office. Their provisions are enforced by the Food and Drug Authorities in each area of the country.

The Milk and Dairies (General) Regulations 1959 cover the registration of dairy farms, dairy farmers, and dairies; the inspection and health of cattle; general provisions relating to buildings and water supplies; and provisions to ensure that the production, treatment, handling, storage, conveyance, and distribution of milk are carried out with adequate safeguards against contamination or infection from unclean surroundings or dirty equipment.

CLASSIFICATION AND DESIGNATION OF MILK

Milk is purchased from the dairy farmer by the Milk Marketing Board. The price the farmer receives for his milk depends on its content of solids-non-fat (SNF) (see p. 3) and total solids. The basic price is paid for milk containing 8·40 per cent or more SNF and 12·4—12·5 per cent total solids; additions are made to the price paid for milk containing more total solids and deductions for milk containing less. There is no legally defined lower limit to the solids content of milk, though under the Food and Drugs Act milk containing less than 3 per cent fat or 8·5 per cent SNF is presumed to have been adulterated unless it can be proved that it is genuine and that nothing has been added or removed since it left the cow. Regulations also require that all milk for human consumption sold as Channel Island, Jersey, Guernsey, or South Devon milk shall contain not less than 4·0 per cent fat.

The Milk Marketing Board acts as a wholesaler and sells the milk to dairy firms which process and distribute it. The price that can be charged to the consumer is laid down by Government order.

Nearly all milk sold by retail in England and Wales must be specially designated, as laid down by the Milk (Special Designation) Regulations 1963 and the Milk (Special Designation) (Amendment) Regulations 1965 and 1972. There are four categories: untreated, pasteurized, sterilized, and ultra-heat-treated (UHT). Untreated milk is raw milk which has not been heat-treated and which has been bottled on the farm or at the dairy. The regulations lay down the heat-treatments to be used for pasteurizing, sterilizing, and UHT processes, and give details of three chemical tests by which the effectiveness of the procedures can be checked.

1. The *methylene blue test* gives an indication of the number of bacteria in untreated or pasteurized milk. Methylene blue is a blue dye that is rendered colourless by bacteria. In the test methylene blue is added to milk and the test is passed if the blue colour is not lost within 30 minutes.

2. The *phosphatase test* shows whether the proper temperature has been reached during pasteurization. Heat destroys the milk enzyme phosphatase, so that if the test shows the presence of this enzyme, the milk must have been improperly pasteurized.

3. The *turbidity test* for in-bottle sterilized milk is based on the finding that the whey proteins of milk are completely denatured during proper sterilizing. The test is carried out by adding ammonium sulphate to milk, filtering off the resulting precipitate and then heating the filtrate. The filtrate from untreated milk and from pasteurized, boiled, or UHT sterilized milks gives a turbid solution on heating, due to coagulation of the undenatured whey proteins present. The filtrate from in-bottle sterilized milk should contain no undenatured whey proteins and should give no turbidity on heating.

The procedures used to carry out these tests in the laboratory are described in the Appendix (p. 60).

REGULATIONS CONTROLLING THE COMPOSITION AND LABELLING OF MILK PRODUCTS

These include the following

1. The Condensed Milk Regulations 1959, which specify the declarations

that must be carried by labels on containers of condensed and evaporated milk and the minimum contents of milk fat and milk solids that must be present.

2. The Skimmed Milk and Non-milk Fat Regulations 1960 and (Amended) 1968, which are concerned particularly with the composition and labelling of baby foods containing fats other than milk fat.

3. The Dried Milk Regulations 1965, which specify the declarations that must be carried by labels on containers of dried milk powder and the percentage content of milk fat that must be present in the various types of dried milk.

4. The Cheese Regulations 1970, which specify requirements for the composition and description (including labelling) of varieties of hard, soft, whey, and processed cheeses and cheese spread.

5. The Cream Regulations 1970, which specify requirements for the description (including labelling and advertising) and composition of the various types of cream.

6. The Butter Regulations 1966 and the Margarine Regulations 1967, which specify requirements for the composition and labelling of butter and margarine.

7. The Ice Cream (Heat-Treatment, etc.) Regulations 1959 and 1963, which define the allowed methods for pasteurizing or sterilizing the mix before freezing, and the Ice Cream Regulations 1967, which specify requirements for the composition and labelling of ice cream and Parev (kosher) ice.

Conclusion

Milk and dairy foods are an important source of nutrients in the UK, as they are in many other countries. Milk is the most nearly complete single food; the procedures at present used in processing liquid milk and in preparing a wide range of dairy foods are generally conservative, so that in most products the losses of nutritive quality are small.

All those concerned with the production and sale of milk and milk-based foods are hoping for an expansion of the existing demand. If this is to happen it is likely to depend to an increasing extent on the development of new products attractive to the consumer. It is probable that particular attention will be given on the one hand to concentrated products from whole milk that reduce the need for packaging and transporting large volumes of water, and on the other to the preparation and utilization of the individual constituents of milk (fat, protein, and lactose) as raw materials for the manufacture of other foods. For example, attention has recently turned to the use of methods of concentrating milk and whey that do not involve the use of heat; developments in physico-chemical techniques, such as gel filtration and ultrafiltration, have provided methods of producing undenatured whey proteins which have functional properties making them valuable for inclusion in other processed foods. Another interesting new development in dairy products is the preparation of co-precipitates of casein and whey proteins by acidifying heated milk or mixtures of cheese whey and skim milk. Co-precipitates have been developed independently in Australia and New Zealand to provide an effective means of conserving milk proteins in products that also contain all the calcium and phosphorus associated with the casein. The good functional properties of co-precipitates and the high nutritional value of their proteins again make them attractive raw materials in the food industry.

It is to be hoped that as much attention will be given in the future as has been given in the past to ensuring that processing procedures do not impair the nutritional quality of the ingredients of milk or the unique properties of milk itself as a food.

Questions

1. Explain why milk is considered to be such a valuable food. (City and Guilds)
2. Describe the main differences in composition between milks secreted by the Friesian and the Guernsey breeds of cattle. How do these differences affect the nutritive value of the milks?
3. Explain how milk goes sour and suggest ways of preventing this. (City and Guilds)
4. Compare the processes of pasteurization and sterilization of milk. What changes occur in the nutritive value of the milk in each case?
  What is the effect of light on the nutritive value of (a) milk in a clear-glass bottle; (b) milk in an aluminium-foil-lined carton?
5. What are the main differences in the production of sterilized milk and ultra-heat-treated milk? How do these products differ in nutritive value? (City and Guilds)
6. Describe the method of heat treatment used in processing any two of the following:
  (a) pasteurized milk; (b) milk for cheesemaking; (c) ice cream mix; (d) cream for buttermaking. (City and Guilds)
7. Outline the main stages in the manufacture of hard cheese. Give an average composition for hard cheese.
8. Write short notes on any three of the following:
  (a) lactose; (b) dried skim milk; (c) filled milk; (d) soft cheese; (e) whey; (f) single cream.
9. Calculate the energy of (a) a glass of milk (200 ml); (b) a portion of butter (15 g); (c) a serving of cheese (20 g). What approximate percentage contribution would each of these make to the daily protein needs of a 7-year-old schoolboy?
10. How does the composition of human milk differ from that of cow's milk? What must be added to cow's milk to make it a complete food for the human infant?
11. Explain what you understand by the term 'overrun' in the manufacture of ice cream.
  An ice-cream mix contains 60 g of butter, 120 g of sugar, and 116 g of dried skim milk per kg of mix. It is frozen with a 90 per cent overrun. Calculate the energy content of a serving from a large scoop which has a volume of 100 ml.
12. Describe the main differences in composition between butter and margarine. Do these differences affect their nutritive value?

QUESTIONS

13. Write short notes on any three of the following:
    (a) UHT treatment; (b) milk proteins; (c) blue-veined cheese; (d) vitamins in milk; (e) whipping cream; (f) cottage cheese.
14. Describe how you could prepare yoghurt in the home. Compare the nutritive values of natural yoghurt, sweetened yoghurt, and milk.
15. Do you think that the slogan 'Drinka Pinta Milka Day' is a good one? If so, why?

Appendix

THE METHYLENE BLUE TEST

Reagent
One standard methylene blue milk-testing tablet is dissolved in 200 ml of cold, sterile, glass-distilled water in a sterile flask and the solution made up to 800 ml with cold, sterile, glass-distilled water. The solution should be stored in a cold, dark place and discarded after 2 months.

Method
The milk to be tested is mixed thoroughly and a 10 ml sample is measured into a calibrated 150 mm × 16 mm test tube, taking care that one side of the interior of the test tube is not wetted with milk. One millilitre of methylene blue solution is added from a 1 ml blow-out pipette without letting the pipette touch the milk in the tube or the wetted side of the interior of the tube. After waiting for 3 seconds, the solution remaining in the tip of the pipette is blown out. The test tube is then closed with a sterile rubber stopper and inverted twice slowly. It is then placed in a warm water bath maintained at between 98·5 °F and 100·4 °F (37–38 °C).

A control tube is used for comparison with each batch of tubes containing milk under test, to indicate when decolorization is complete. It is prepared by immersing in boiling water for 3 minutes a stoppered test tube containing 1 ml of tap water and 10 ml of milk.

The test samples are regarded as being decolorized when the whole column of milk is decolorized or is decolorized up to within 5 mm of the surface. (A trace of colour at the bottom of the tube may be ignored.)

Result
The test is deemed to be satisfied by milk which fails to decolorize methylene blue in 30 mins.

THE PHOSHATASE TEST

Reagents
1. The buffer solution is prepared by dissolving 3·5 g of anhydrous sodium carbonate and 1·5 g of sodium bicarbonate in distilled water and making up to 1 litre.
2. The buffer-substrate solution is prepared by weighing 0·15 g of disodium *p*-nitrophenyl phosphate into a 100 ml measuring cylinder and making up

to 100 ml with the buffer solution. The solution should be stored in a refrigerator and protected from light. The solution should be discarded if not used within one week.

Method

5 ml of the buffer-substrate solution is placed in a 150 mm × 16 mm test tube fitted with a rubber stopper and placed in a water bath maintained at 99 °F ± 1 °F (37·5 °C ± 0·5 °C). When the temperature of the solution has reached 98·5 °F (37 °C), 1 ml of the milk to be tested is added and the contents of the tube are well mixed by shaking. The test tube is then incubated for 2 hours at 98·5 °F (37 °C). A blank or control tube containing 5 ml of the buffer-substrate solution and 1 ml of boiled milk is incubated with each series of samples.

After incubation each test tube is removed from the water bath and its contents well mixed. The colour of tubes containing milk under test is then compared with that of the blank tube, using a reflected light source and a Lovibond 'all purposes' Comparator fitted with comparator disc A.P.T.W. or A.P.T.W.7.

Result

The test is deemed to be satisfied by milk which gives a reading of 10 μg or less of *p*-nitrophenol per ml of milk.

THE TURBIDITY TEST (for in-bottle sterilized milk, see p. 20)

Method

Four grams of ammonium sulphate is weighed into a 50 ml conical flask, 20 ml of the milk sample added, and the flask shaken for 1 min to ensure that the ammonium sulphate dissolves. The mixture is left for 5 min and then filtered, through a folded 12·5 cm No. 2V Whatman filter paper, into a test tube. When 5 ml of a clear filtrate have collected, the tube is placed in a beaker of boiling water for 5 min. The tube is then transferred to a beaker of cold water and, when the tube is cool, the contents are examined for turbidity by holding it in front of an electric light shaded from the eyes of the observer.

Result

The test is deemed to be satisfied when the sample of milk shows no sign of turbidity.

Further reading

Cheke, V. and Sheppard, A. (1965). *Cheese and butter.* Hart-Davis, London.
Davidson, Sir S, and Passmore, R. (1963). *Human nutrition and dietetics.* E. & S. Livingstone Ltd., London.
Davis, J. G. (1955). *Dictionary of dairying.* Leonard Hill, London.
− (1965). *Dictionary of dairying* (supplement to 2nd edn). Leonard Hill London.
− (1965−7). *Cheese,* Vols I and II. J. A. Churchill Ltd., London.
Hyde, K. A. and Rothwell, J. (1973). *Ice cream.* Churchill Livingstone, New York.
Kon, S. K. (1972). *Milk and milk products in human nutrition* (2nd edn). Nutritional Studies no. 27. Food and Agriculture Organization.
Lampert, L. M. (1970). *Modern dairy products.* Food Trade Press Ltd., London.
Layton, T. A. (1967). *Cheese and cheese cookery* (The Wine and Food Society). Michael Joseph, London.
Ministry of Agriculture, Fisheries and Food (1968). *Manual of Nutrition.* H.M.S.O.
Webb, B. H. and Johnson, A. H. (1974). *Fundamentals of dairy chemistry.* Westport, Connecticut, U.S.A.
Widdowson, E. M. and McCance, R. A. (1967). *The composition of foods.* H.M.S.O.

Index

acetic acid, 4
amino acids, essential, 9, 15, 24, 27, 39
antibodies, colostral, 6, 29
ascorbic acid, 12, 17, 18, 20, 21, 31

biological value, protein, 9, 27
blue-veined cheese, 36
boiled milk, 18
breeds of cow, 5
buffalo's milk, 6
bulk milk, 2
buttermilk, 53
B vitamins, 12, 13, 17, 18, 20, 21, 23, 27, 38, 53
   synthesis in rumen, 5
browning, non-enzymic, 27

calcium, 10, 13, 31, 39
carotene, 1, 5, 6, 11
casein, 2, 9, 31, 39, 57
   biological value, 9
cellulose digestion, 4
Channel Island milk, 5, 55
Cheddar cheese, 34, 39
cheddaring process, 34
cheese
   keeping quality, 34−8
   varieties, 35−9
clotted cream, 42
colostrum,
   cow, 6
   human, 29
concentrated milk, 22, 57
condensed milk, 23
   sweetened, 22
co-precipitate, 56
Cornish cream, 42
cottage cheese, 37, 39
cow breeds, 5
cream cheese, 37
cream, 42−3
   home-made, 43
cultured buttermilk, 53

Designated milk, 55
Devonshire cream, 42
diacetyl, 46
double cream, 42

evaporated milk, 22
ewe's milk, 6

fatty acids, 8
   polyunsaturated, 8, 46
filled
   cheese, 38
   milk, 22
folic acid, 13, 20, 21, 23
food value, *see* nutritive value
Friesian cow, 5

goat's milk, 6
Guernsey cow, 5

hard
   cheese, 34, 39
   ice cream, 47
homogenized milk, 19
human milk, composition of, 29

ice cream
   hard, 47
   soft, 48
iced lollies, 47
in-bottle sterilized milk, 20
instant milk, 26
iron, 10, 31

keeping quality,
   cheese, 34−8
   milk, 18
      concentrated, 23
      dried, 27
      UHT sterilized, 20
kefir, 51
kumiss, 51

α-lactalbumin, 2
Lactobacillus bulgaricus, 51
β-lactoglobulin, 2
lactose, 3, 7, 40
   intolerance, 7
lipase, 19,
lysine, 15, 24, 27, 39

Maillard reaction, 27
methylene blue test, 55, 60

milk,
  consumption, 13
  fat, synthesis, 3, 8
  human, 30
  keeping quality, 18
  loaf, 10
  Marketing Board, 54
  proteins, 2, 9, 14
  sugar, 3, 7
  vitamins, 3
  yield, 1, 5
minerals, 10

National dried milk, 31
non-enzymic browning, 27
non-fatty solids, 2, 4, 5, 54
nutrient requirements, child, 14
nutritive value,
  butter, 44
  cheese, 39
  cream, 43
  ice cream, 49
  margarine, 46
  milk,
    concentrated, 23
    dried, 26
    in-bottle sterilized, 21
    pasteurized, 18
    UHT sterilized, 20
  yoghurt, 52
overrun, ice cream, 48

pasteurized milk, 17
Penicillium
  candidum, 36
  roquefortii, 36
phosphatase test, 55, 60
polyunsaturated fatty acids, 8, 46
propionic acid, 4
processed cheese, 37
protein,
  amino-acid composition, 9
  biological value, 9, 27
  complementary, 9, 15, 40

retinol, 1, 5, 6, 11, 13, 31, 45
riboflavin, 12, 13, 17, 38, 53
rind, cheese, 36, 39
roller-drying, 25
ruminants, 4

School Milk schemes, 13
semi-hard cheese, 34, 39
sherbet, 49
single cream, 42
skim milk, 22
soft
  cheese, 36, 39
  ice cream, 48
solids-not-fat (SNF), 2, 4, 5, 54
sorbet, 47
sour
  milk, 16
    fermented, 51
spray-drying, 26
standardized milk, 21
sterilized
  cream, 42
  milk, 19
Streptococcus
  cremoris, 34, 37
  lactis, 34, 37
  thermophilus, 51
sweetened condensed milk, 22

toned milk, 22
thiamin, 12, 13, 18, 21, 23, 38, 53
triglycerides, 2, 7, 45
turbidity test, 55, 61

UHT sterilized milk, 19

vegetable-fat cheese, 38
vitamin
  A, 1, 5, 6, 11, 13, 31, 45
  B, 12, 13, 18, 21, 23, 38, 53
  B_{12}, 13, 18, 20, 23, 27
  C, 12, 17, 18, 20, 21, 31
  D, 1, 11, 23, 31, 45
vitamins in milk, 3

water ice, 47
whey, 2, 38, 40
  cheese, 37
  proteins, 2, 31, 40, 55, 57
    biological value, 9
whipping cream, 42

yoghurt, 51
  home-made, 52